CALL OF NATURE

CALL OF NATURE

THE SECRET LIFE OF DUNG

RICHARD JONES

Pelagic Publishing | www.pelagicpublishing.com

Published by Pelagic Publishing
www.pelagicpublishing.com
PO Box 725, Exeter EX1 9QU, UK

Call of Nature: The Secret Life of Dung

ISBN 978-1-78427-105-3 (Hbk)
ISBN 978-1-78427-106-0 (ePub)
ISBN 978-1-78427-107-7 (Mobi)
ISBN 978-1-78427-108-4 (PDF)

A catalogue record for this book is available from the British Library.

Printed and bound in Great Britain by Clays Ltd, St Ives plc

Cover images:
Dung beetle (Scarabaeinae sp.) by Chris Shields
African elephant by Vaclav Volrab (Shutterstock)

CONTENTS

PREFACE

We don't talk about it much in polite society; it's something we do quietly, alone, discretely, behind closed doors. But we all do it. So does the entire animal kingdom. Contrary to popular opinion, excrement is not a dirty word. From a biological perspective, defecation is a fascinating process, and its importance does not stop at the point of voidance.

The fall of dung onto the ground is just the beginning of a complex process of reuse and recycling, and comes with its own intricate ecological web as the multiple dung-feeders and scavengers compete with each other, with predators and parasites, and against the clock to make the best use of the limited quantity deposited in each pat.

The ancient Egyptians were sufficiently enamoured with dung-roller scarabs to create a sun-roller deity in their image. Four thousand years ago scarab amulets were the most popular form of personal jewellery, with representations sumptuously crafted in exquisite and ornate style, or in crude but charming rustic simplicity. We can only guess at their motivation, but it seems pretty clear that the Egyptians put aside any squeamishness about the insects' habits, and celebrated, instead, their ingenuity, their tenacity or their environmentally friendly recycling behaviour.

Without the unsung heroes of the dung fauna we'd soon be knee deep in our own ordure, and that of our farm animals. This is not hyperbole; it very nearly happened on the other side of the world, when the British colonised Australia and took with them their cows and sheep and horses. They made the big mistake of taking familiar grazing animals to an unfamiliar continent. It took 200 years before they thought to take the dung clear-up brigade. They're still a long way from sorting it all out.

Ecology, the interconnectedness of every living thing, is complex beyond any simple measure. We cannot study everything, everywhere, all the time; but we can draw some understanding from looking at the small parts of the world, and seeing how the individual cogs whirr together. This, at least, gives us a sense of awe in the diversity of living organisms and the mind-numbing complexity of our planet. A dung pat is a small, compact, discrete unit, but by watching the comings and goings of the beetles, flies and other animals that recycle it, we can begin to see at least some of that bigger picture.

Dung, then, is the hook on which to hang a series of ecological messages, some bizarre, some astonishing, some actually quite beautiful. There is no need to avert eyes, or turn up noses. It is just one small part of the turning of the natural world. I consider myself very lucky to have been given an interest in natural history when I was a very small boy. And despite all the obvious schoolboy jokes, dung was there early on too.

* * *

My father made eye contact and said something along the lines: 'Can you hear that?' At first I wasn't sure whether he meant Radio 4 droning away in the corner of the room, my brother careening down the stairs, or the kettle whistling on the gas in the kitchen. No, he was referring to an almost inaudible 'tick, tick, tick' coming from the window. He had the knowing look of someone who is about to show off something new.

As a 10 year old, it was not unusual for me to be sitting in the lounge, as we called my father's book-lined sitting room. Whilst he sat in the centre of the room behind the large polished wooden desk strewn with pens, papers and books, perhaps a microscope and a drawer of insects, I'd be perched at the smaller bureau-style table against the wall. Maybe I'd be doing homework. Actually, I'm not sure 10 year olds had homework then. More likely I'd be writing up my own nature diary from whatever family trek we'd been out on that day. I might even have been pinning my own insect specimens, or doodling a sketch of a plant, or a map.

The tapping was definitely coming from outside the window. We drew back the curtains, but the brightly lit aura of the room barely penetrated the dark outside. There was nothing I could see. My Dad knew better. Slipping on shoes we tripped round to the front of the house to see what was going on.

The noise had stopped when we got to the window, but Dad pointed to the windowsill, probably just at or above the level of my eyeline. There, crawling across the yellow paintwork was a beetle. Medium-sized (12 mm), elongate, parallel-sided, subcylindrical, dark brown nearly black, it had shortish stout legs and strongly clubbed antennae – *Aphodius rufipes* was my first dung beetle. It had flown in from the flood-plain grazing meadows that flanked the River Ouse, between our house and the port town of Newhaven across the fields. It had probably come many hundreds of metres, quite an achievement for a half-inch insect.

I strain now, but I can't quite remember whether I thought that living in dung was an odd thing for a beetle. Maybe the notion of dung recycling had already crossed my radar. I certainly understood about stag beetle larvae living in rotten wood. I probably knew about drone flies breeding in flooded tree holes. It's all decaying organic matter.

It wouldn't be long before Dad would also show me the huge dumbledors, *Geotrupes spiniger*, or maybe it was *stercorarius*, heaving its juggernaut way through the fingers of my clasped hand, then flying off, like a miniature helicopter. The power of the toothed legs amazed me, and the feeling of that downdraft as it buzzed away stays with me still.

Dissecting a cow pat came naturally to me. Other dung beetles followed. The great glossy *Aphodius fossor*, slightly shorter, but thicker and heavier than *rufipes*, was a favourite, so too was the small mottled and rather rare *Aphodius paykulli*. The chunky earthmover shape of *Onthophagus coenobita* appeared when I graduated to dog dung, and the mythically horned minotaur beetle, *Typhoeus typhaeus*, was eventually dug up from under rabbit crottels in Ashdown Forest.

I still find *Aphodius rufipes* occasionally, in cow or horse droppings – never at my lighted window though. But whenever I hold its smooth, elegant shape in my fingers, I still think back to the warm summer Newhaven evenings, and the delicate head banging on the lounge glass.

Richard Jones
London, January 2016

ACKNOWLEDGEMENTS

As usual, many people have contributed ideas, knowledge and sage advice to this project. Lillian Ure-Jones was my editor and sounding board, and she read through the entire typescript, spotting my silly spelling slips, pointing out where I had rambled on too long, and offering serious linguistic advice for my glossary. Verity Ure-Jones used her superb drawing skills to provide many of the illustrations that I was unable to find elsewhere, particularly the droppings drawings used in chapter 11, and some of the more obscure dung beetles.

Others have lent me books, sent offprints or copies of articles, and generally chivvied me along. My grateful thanks go to: Robert Angus, Ralph Atherton, Max Barclay, David Buckingham, Roger Booth, Jo Cartmell, Andy Chick, Matthew Cobb, John Cole, Martin Collier, Michael Darby, Malcolm Davidson, Jonty Denton, Mark Depienne, John Drewett, Rosie Earwaker, Garth Foster, Maria Fremlin, Geoff Hancock, Peter Hodge, Jens Horspestad, Sally Huband, Stephen Hutton, Trevor James, A. Jaszlics, Martin Jenner, Jim Jobe, Kate Long, Darren Mann, Paul Manning, Ian McClenaghan, Stephen McCormack, Mike Morris, Nick Onslow, Alan Outen, Hugh Pearson, Bruce Philp, Elizabeth Platt, Nancy Reed, Matt Smith, Sally-Ann Spence, Don Stenhouse, Malcolm Storey, Melanie Warren, Clive Washington, Nichola Whitehouse and Richard Wright.

CHAPTER 1

INTRODUCTION – WHAT IS DUNG?

What's brown and sounds like a bell?
Dung!
Monty Python's Flying Circus, December 1969

IN ITS MOST familiar sense, either from our own personal first-hand experience, or from our close historical proximity to stock animals, dung *is* brown. This is the default colour of the scatological cartoon, or the plastic doggy poop bought in a joke shop. But anyone who has trodden the minefield of dog mess along an urban street knows that the droppings left behind by lazy and careless owners can be anything from pale yellow to red to black. And in widening the boundaries to cover all animal excrement, even this colour palette soon broadens out. Hyena dung is white, bird splashes are piebald, white and grey; reptile waste is anything from pale grey to inky blue-black (my pet garter snake Bella shunts out vaguely greenish goo twice a month); aphid honeydew is clear or, as its name suggests, slightly honey-coloured; caterpillar frass can be black, green or even turquoise verging on blue. Like many things in nature, colour is not a useful guide.

Instead, a better start can be made by trying to understand dung in terms of a simple schoolbook equation of its basic biochemistry:

Food – nutrition + waste = dung

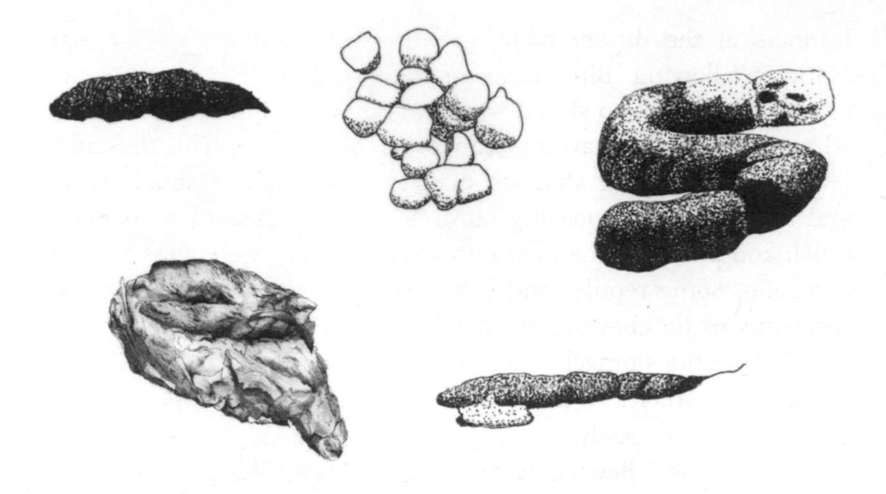

Fig. 1 From black (hedgehog) and white (hyena) through green (goose) and blue-grey (snake), or smelling of violets (otter), dung packages come in all shapes and sizes.

Anyone who can remember back to biology lessons may recall vague snippets about salivary amylase, gastric acid or the pyloric sphincter. Whatever the complex chemistry going on in the intestinal tract, the process that takes food and makes excrement begins with digestion, and to fully appreciate exactly what dung is, it is as well to start with a brief look at this process.

Human digestion is fairly well studied, and since humans are omnivores, eating a huge range of different foodstuffs around the world, our understanding of how it all works is useful when looking at other animals, in particular those mammals with whose similarly brown faeces we are intimately familiar.

PAYING LIP SERVICE TO FOOD

It all starts with chewing, the obvious mechanical breaking down of big bits into smaller bits using the teeth. This makes swallowing a lot easier, but it also helps get the digestive juices working quickly on the ground-up mush, rather than having to cope with rough, tough chunks. Not all animals chew quite as politely as would-be humans at the dinner table, grinding the requisite 40 chomps before swallowing. Birds have no teeth, and apart from a cursory crunching, usually to stop it wriggling, they gollop down their food whole. Instead, they have a gizzard to do the chewing. This muscular portion of the upper stomach often contains grit or small stones and the rhythmic squeezing movement of the gizzard walls helps crush and pulverise the food into a more manageable substrate for digestion. Some reptiles and fish also have gizzards. Whatever the mechanisms for chewing, the result is a more-readily digestible raw material for the stomach – it's all about increasing the surface area of the food particles so that the chemical processes of digestion can get to work more easily.

In humans, chewing is not just about cutting and crushing; digestion of some foods actually begins in the mouth, with digestive enzymes in the saliva. Starch (the primary carbohydrate in foods such as bread, potatoes, pasta and the like) is attacked by salivary amylase to produce various sugars. A simple home experiment involves over-chewing a piece of bread (without swallowing), which becomes noticeably sweeter after only a couple of minutes. My old biology teacher Mr McCausland introduced me to that one during O-levels (the old name for GCSEs, for any younger readers), ostensibly to show us a practical demonstration of hydrolysing starch catalysis, but I suspect also to fill our incessantly noisy mouths with slightly stale loaf to shut us up for a bit. The mouth is a neutral or slightly alkaline environment, but the swallow takes boluses of chewed food down the oesophagus (gullet), into the highly acid stomach.

Although the hydrochloric acid in the stomach is strong enough to dissolve iron, its purpose is not just to attack the food, but rather to create the right chemical environment in which highly complex food-digesting enzymes can get to work. Serious protein digestion gets going now as these immensely complicated molecules are snipped into smaller units. The acid also kills most bacteria should any have been on the food when it was eaten. Eventually a sloppy 'chyme' is produced – this is the smelly pale-yellowish acrid liquid full of chopped carrots which is revealed if you are unfortunate enough to vomit a couple of hours after eating. Chyme is slowly released in gentle squirts through the pyloric sphincter, the muscle-ring one-way valve at the far end of the stomach, down into the small intestine.

The human small intestine is another world altogether. A convoluted 6 metre stretch of narrow, finger-thick tubing, it is the major site of digestion and food absorption. Several profound chemical changes start within the first few centimetres. Sodium bicarbonate is released by the pancreas, a large gland sitting under the stomach; this neutralises the stomach acid and creates a slightly alkaline background. The pancreas also releases alkali-controlled enzymes – important components are proteases and peptidases to continue the digestion of proteins, and more amylase to break down starches. A thick yellowish-brown liquor called bile is also dribbled into the intestine from the gall bladder. It helps break down clods of insoluble fat into an emulsion of microscopic globules. The bile also contains a yellow waste substance called bilirubin, which is made from haemoglobin (the red oxygen-carrying molecule in the blood) as damaged red blood cells are broken down and destroyed in the liver. As it traverses the digestive tract, bilirubin changes to another strongly coloured chemical called stercobilin, it is this dark brown pigment that gives mammalian dung its characteristic brown colour.

YOU ARE WHAT YOU EAT

Most of the bewilderingly sophisticated biochemical substances that make up living organisms, and therefore the food that we eat, are based upon long chains of repeating chemical units, like the beads on a series of pearl necklaces. The chains fold and twist, and are cross-linked like knitting to form everything from the proteins which build the bulk of muscles, and the meat that we eat, to starch (the energy source in foods such as spaghetti and doughnuts) and polyunsaturated fats (our consumption of which we try to reduce by choosing low-fat margarines). The enzymes from the pancreas, and others secreted by the small intestine itself, work on digesting our food like chemical scissors, trimming, then snipping off individual beads from these intricate long-chain necklaces. Truly vast substances containing many thousands (sometimes millions) of atoms are reduced to the basic tiny molecules of the individual units from which they are made up.

Proteins are reduced to their constituent amino acids, starches are reduced to simple sugars, and fats are reduced to short-chain oils. Each of these fundamental building block types is just a few atoms (maybe 10–100) in size – and small enough to pass through the semi-permeable membranes of the gut wall, to be whisked off around the body in the blood and lymphatic transport systems. To facilitate this removal of useful chemical nutrients from the chyme, the interior surface of the small intestine is wrinkled and minutely convoluted, covered all over with tiny finger-like extensions (villi) that give it the appearance under the microscope of a thick shag-pile carpet. This has the effect of hugely increasing the surface area across which nutrient absorption into the blood takes place. Choose your own incredible statistic here, the usual one is that the surface area of the human small intestine, if it were to be flattened out, is the size of a tennis court (260 square metres). If memory serves, that same Mr McCausland (in A-level biology now) had us calculating this based on a section of sheep intestine he'd got from his neighbourhood abattoir. After poring over the microscope and calibrating the graticule eyepiece measurement scale, we probably spent the entire lesson counting sheep villi and extrapolating up from microscopic cylindrical tendrils to so-many hundred square metres of laboratory floor carpet.

The final leg of the digestive conveyor belt is the large intestine, the colon, about a metre long and wrist-thick. Some last-minute nutrient absorption occurs here, but its most important function is to remove water from the digestion remains.

By now the semi-liquid chyme has become a stiff semi-solid. After, perhaps, several days of slow onward movement through the digestive tract, much of what the human body can use in the way of nutrients has been removed from the food. What remains is the key constituent of plant and vegetable foodstuff that we cannot digest – fibre (sometimes called roughage in older textbooks). Fibre is made up of undigestible chemical chains such as cellulose and lignin. These are the substances that give plants their incredible toughness and strength, and which in non-food species provide us with fibres for other uses – cotton jeans, linen sheets, wood-pulp paper.

Fibre was a major preoccupation for my parents' generation, and whereas I was fascinated by the numerous helpful vitamins thoughtfully provided and carefully listed on the packets by the manufacturers of sugary breakfast cereals, my Mum's shopping choices were more often influenced by the roughage content of the bran-based stodgier end of the edibility spectrum. It was during the 1950s and 1960s, when processed foods, cheaper meat and sliced white-pap bread started to appear on the supermarket shelves, that the connection between healthy food intake and healthy stool output was first used (albeit very tastefully and subtly) as a marketing tool. Being 'regular' was considered a feminine virtue and a sign of manly rectitude. There was, at this time, a growing realisation that the human taste for sweet, easily digestible titbits was replacing a more rounded diet of mixed fibrous fruit and vegetables, and that this was playing havoc with a digestive tract inherited from our long-distant ancestors, one more suited to the foraged nuts, fruit and roots on which proto-humans first fed.

Lack of fibre in the diet didn't just mean reduced faeces; this was not just a simple equation of less in one end, less out the other. It meant less regular throughput, intestinal stagnation, backing up of waste, and rectal compaction to the point of discomfort and risk to health. Constipation is a singularly human obsession, and we'll be straining to understand its implications regularly throughout this book.

By lucky happenstance, I was able to do some personal research into constipation early on in the writing of this book, when I was hospitalised with extreme abdominal pain and excruciating muscle cramps. Fearing it might be gall or kidney stones, hernia or diverticulitis, I trekked up to my local accident and emergency unit late one Saturday evening. I was poked and prodded, drained of numerous blood samples and eventually X-rayed, but the tests were negative and the diagrams showed it was, to give it the medical term, simply faecal loading. I was a bit bunged up in there. Several days of laxative oils and glycerine suppositories got things moving again, much to everyone's relief.

There is another key ingredient in the final quasi-excrement as it passes through the large intestine, and by the time it is ready to be ejected from the body, it makes up over half the dry weight of the faeces – bacteria. This is where our knowledge of human digestion starts to wear a bit thin. There may be 100 trillion bacteria in an average human intestine, that's 100 million million, or 1 with 14 zeros after it – a mind-boggling number, and more than ten times the number of body cells in the average human's entire body. There are thought to be 300–1,000 different bacterial species in there; it's difficult to quantify exactly, because they are difficult to identify and difficult to grow and study in laboratory cultures. What are they all doing?

Most people's idea of 'bacteria' may be of horrible germs that cause sickness, disease or death, but the gut flora, to give it its usual, slightly more passive, euphemistic name, is a perfectly normal, healthy, indeed necessary part of being a human. Unlike the bacteria that cause, say, tuberculosis, cholera or salmonella poisoning, these natural gut-dwelling microbes are not attacking or parasitising the human body, nor are they accidental inhabitants (sometimes called commensal, meaning non-harmful coexistence), they are better described as being mutualistic – host and occupier each benefiting from the presence of the other.

The bacteria benefit because they are supplied daily with a fresh input of partly digested food passing through the guts, on which they and their descendants can feed; and in return they further digest the remaining substances, their own very different enzymes snipping away at the slowly fermenting chemical dross that would otherwise be unavailable to our own bodies' somewhat limited digestive machinery. These last-minute digestive products are absorbed, along with some of the water, before the final bodily waste product is ready to be voided.

This usual, normal, natural process goes on in the human body every day, and for the most part we are completely oblivious of it, but when things change, we can get a fascinating insight into the digestive processes, and we can understand parallels in other animals.

RIGHTS AND WRONGS OF PASSAGE

Human stool is roughly 75% water. Anyone with a regular balanced diet should be familiar enough with their own bowel movements to know when things are 'fine down there', but when things go wrong, either way, we notice immediately. Water content in human faeces can actually range from about 50 to over 90%; this translates to an ease-of-passage range, from reluctant constipation to explosive diarrhoea.

Helpfully, there is a simple medical scale, with pictures – the Bristol stool chart – which classifies this range into seven discrete categories by outward visual appearance. No messy weight and density tests are necessary.

At the one extreme, diarrhoea can be life-threatening; the International Centre for Diarrhoeal Disease Research in Dhaka, Bangladesh, was set up because, after malaria, diarrhoea is the world's single biggest killer of children under 5. The usual cause is viral or bacterial infection of the gut by inappropriate micro-organisms, and the body's response is to flush out the system as quickly and effectively as possible to get rid of the offending invaders. Shutting down or reversing the intestines' water-absorption pathways keeps the gut contents highly liquid, and the rhythmic waves of muscular contraction (peristalsis) that usually gently squeeze the digesting food through the digestive tract now force it out under pressure. Fast ejection (like vomiting) gets rid of the offending microbes and helps prevent dangerous, possibly life-threatening bacterial toxins building up; it may even stop these micro-organisms invading the interior of the body.

We all have our own diarrhoeal anecdotes, and in any other circumstances I'd keep mine diplomatically quiet. But since this is actually a book about my own exploration of excrement I cannot pass without at least commenting obliquely on the accident in the Temple of the Buddha's Tooth in Kandy, Sri Lanka, in 1992, where it was forcefully brought home to me that not all of the world's drinking water is safely potable. I was saved major embarrassment by the hasty hailing of a friendly tuk-tuk driver who quickly returned me to the safe and private confines of the family-run guest house in which we were lodging, where for several days I closeted myself away in my room and recuperated on thin vegetable gruel and bottled water.

Diarrhoea can also be caused by allergic reactions, physical or chemical damage to the gut linings, poisoning, alcohol abuse or age-related blood-vessel damage. As well as any underlying cause, the primary health risk, especially in the very young or the very old, is dehydration because of the body's continued and copious water loss. The usual medical response is treatment with rehydrating fluids (thin soup is ideal) carefully balanced to replace sugars and salts that also pass from the body during diarrhoeal attack.

Fig. 2 Charming rural vignette from Bewick's *A General History of Quadrupeds* (1790) with obligatory pat.

Any visitor to a dairy farm may wonder if cows suffer constant diarrhoea, since their excrement is very runny and forms splashed circular pats when it pours messily out. For cows, however, this liquid dung is entirely normal, because rich green grass is a natural laxative (it has a very high fibre and water content) and bovine nutrition is based on the bucket digestion technique rather than the long narrow tube system used by humans.

The cow's digestive system works on a much larger scale than a human's, not just because the cow is physically larger, but because it is specially adapted to cope with processing enormous quantities of tough grass cellulose. Initial grazing is little more than cutting and swallowing; the chewing is done later, at the cow's convenience. A large and multiple-compartment stomach allows the cow to regurgitate mouthfuls of 'cud' from the first portion of its stomach, the reticulum, which acts as a large storage pouch. Grinding of the grass occurs as the cow ruminates (chews the cud), and often looks like the cow is insolently chewing gum as it nonchalantly lounges about in a sheltered portion of the field. The second time the food is swallowed it passes into the largest section of the stomach, the rumen, often vibrantly described in agricultural information leaflets as being the size of a garden dustbin; here bacterial-led bucket-style fermentation really gets to grip with digesting the tough grass cellulose. Eventually the fibrous soup passes through small, then large intestines, with their concomitant nutrient and water absorption.

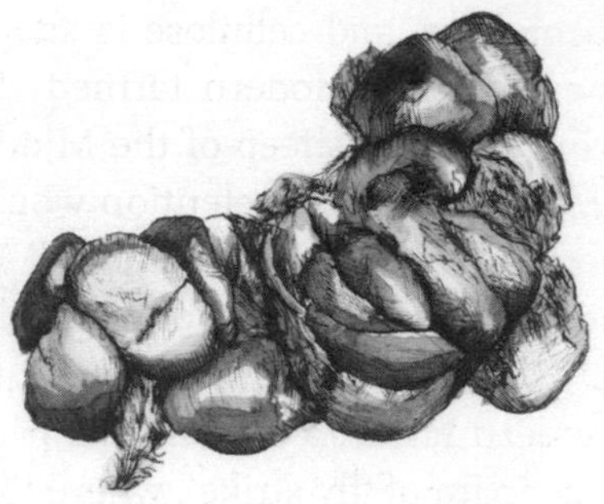

Fig. 3 A cow pat is roughly 75% water, a semi-liquid which pools into neat round field ornaments. Horse dung is only slightly less water (72%), but is firmer, with longer strands of fibre, so maintains its shape as decorative road apples.

The water content of cow dung is roughly 75%, remarkably similar to that of normal human faeces. The liquid nature of cow dung is down to the fact that the fibre (which is mostly undigested in humans, and so gives our stools their firm texture) is much more broken down by the cow's stomach bin of bacterial enzymes, and it exits the animal in a more fluid, rather than a semi-solid state.

Despite the fact that horses also graze grass, their droppings are not at all liquid, indeed their dry, rather pleasantly aromatic dung sometimes seems little more than the partially decomposed grass cuttings on the compost heap, making it the manure of choice for gardeners and smallholders. Horses also digest grass via bacterial fermentation, but instead of this taking place early on in the gastro-intestinal tract, in the stomach, it takes place much further down the digestion pathway, in the caecum, a pouch at the junction of small and large intestines. The caecum is reduced to the appendix in non-grass-eating humans, but in horses it fulfils its original digestive purpose, and is a metre-long sack.

Horse dung is also about 72% water, just a little drier than cow or human dung, but perhaps the absence of cud chewing and late digestion allows more complex fibre to pass through into their droppings. If their digestion is slightly less efficient than that of cows, horses make up for it by spending more time actually grazing, and less time chewing the cud.

Sheep dung is drier still, 65% water. Sheep, like cows, are

ruminants, and cellulose is attacked in the large rumen portion of the stomach. Modern farmed sheep are thought to be descended from the wild sheep of the Middle East, traditionally a hot and arid area, where water retention would have been important for survival before they were domesticated. Passing copious liquid dung would never have been an evolutionarily successful strategy. It would also have clogged their woolly hindquarters, attracting flies; in well-run modern farms, a sick animal passing loose stools can still easily be the victim of 'fly-strike', where fly maggots infest the dirtied wool on the rump, and even start to attack the animal's underlying flesh.

An added bonus of dry sheep dung is that it makes it much more convenient to handle when looking for dung beetles. There was that occasion, I remember it well, crossing a sheep-grazed meadow on the South Wales coast, where I insisted on picking up (with my bare hands) and breaking open the hard semi-dry sheep nodules, to find various western and montane dung beetles I had never seen before. It was all very exciting, but it wasn't long before I was told by my girlfriend, in no uncertain terms, that if I did not stop forthwith, our relationship would be at an end. I had to be a bit more discreet after that.

Rabbits are also grass-grazers and have evolved a third mechanism for releasing the nutrients tied up in the tough cellulose fibres. Like the sheep, they have a caecum at the junction of small and large intestines, where bacterial fermentation takes place, but, perhaps because of their small size, this does not always extract enough nutrition for them. Instead, they recycle the food, by eating their own dung.

As they feed, rabbits release hard, dry, round pellets (crottels), pale brown like compacted hay, the size of children's small marbles; these are the final waste droppings and the familiar, almost odour-free residue left on their grazing grounds, on the ant hills from which they scan for danger, or in the domestic hutch. But as the day's new intake of grass reaches the end of the digestive tract on its first pass through, rabbits also produce dark, smooth, soft, mucus-covered pellets called caecotropes. These are re-eaten (caecotrophy) as they are produced, and are taken directly into the mouth, from the rabbit's anal vent, usually in the safety of the burrow; they are sometimes also called 'night faeces', so the behaviour is seldom directly observed.

Some friends of mine once kept a rabbit as a house-pet, but they were less than impressed to observe this behaviour close-up, at night, on the pillow, as the friendly rabbit visited them whilst they tried to sleep.

The mucus covering allows the caecotropes to pass through the acid of the stomach, so that nutrients can be extracted from the food by continued bacterial fermentation during its second pass through the gut.

Most other mammal grazers, whether of grass, herbs or tree leaves, use variations on these digestive mechanisms to extract the sometimes meagre nutritive value from readily available but poor-quality plant foodstuffs; and they produce dungs of relatively similar chemical and fibre make-up, although differing in their water content and size according to each species' evolutionary history and individual ecology.

THE LONG AND THE SHORT OF IT

As a general rule carnivores have much shorter intestines, relative to their outward body size, than do herbivores. They do not need the lengthy processes that herbivores require to digest those chemically tough plant fibres such as cellulose. Digesting a meat meal takes much less time and energy than digesting vegetables.

Typical hunting carnivores – such as cats, dogs, foxes, hyenas, wolves – have very short gastrointestinal tracts. Partly this is because proteins are quickly and efficiently metabolised by the well-oiled digestive enzymes in the stomach and intestines, but partly this is a response to the quality of the meat meals that these animals actually get to eat.

There runs an old adage, fondly regurgitated by biology lecturers:

> Q: What does a herbivore eat?
> A: Herbs.
> Q: What does a carnivore eat?
> A: Whatever it can get.

Despite graphic images from wildlife documentaries showing

Fig. 4 Another vignette from Bewick's *A General History of Quadrupeds* (1790), which was intended as a children's book.

dramatic hunting kills by big cats or wolf packs, being a carnivore is fraught with difficulties. Hunting other animals is a dangerous business, where prey is likely to fight back as if its life depended on it – with the very real potential of serious injury to the predator. Chasing large mobile prey can be dangerous and exhausting. There is always a balance between hunting easy small prey, sometimes barely a mouthful, versus bringing down a prize trophy to feed the victor to satiation, and perhaps all of its pack too. This leads to some less than savoury food choices being made by the hungry fox or the desperate lioness. Scavenging from the kills of others, from the victims of disease, starvation or thirst, or making do with nutritionally dubious worms, maggots, caterpillars and other insects, means that the digestive systems of many top-end predators are subject to the sort of chemical or bacterial insult that would cause serious, even fatal, food poisoning in humans.

The shortened digestive tract means that such rancid or infected food passes swiftly through the gut, allowing at least some nutritive goodness to be removed, but ejecting the remains before mortally dangerous bacterial toxins can build up. Finding a small pool of strong-smelling liquid fox diarrhoea on the patio by my back door is sometimes a sign that the animal has been scavenging rotting food from the bins again.

Mammals might produce the dung with which humans are most familiar, but they are in the minority when it comes to the general mechanisms of expelling bodily waste. Mammals produce solid excrement from the anus, but they also produce a separate liquid in the form of urine. Urine is produced by filtration of the blood passing through the kidneys. It is mostly water (90–98%), but also contains waste products from the body's daily activities, notably a substance called urea – $CO(NH_2)_2$. When the body assembles or disassembles amino acids (those digested chemicals broken down from proteins in food) into its own complex biochemicals, there is often an excess of nitrogen (N) compounds; the simplest chemical reaction would produce ammonia (NH_3), but this is highly toxic to most organisms, hence its use in powerful domestic cleaning products. Urea combines two ammonium units into a substance that is neither acid nor alkali, is highly soluble in water and relatively non-toxic. Although urine is not pleasantly appetizing, it is also at least not revoltingly undrinkable; otherwise all those dreadful survival tales would never have had survivors to tell them. Urine can be safely stored in the bladder until it is convenient to expel it.

In mammals, urine and faeces are discharged separately, but in almost all other living creatures, from albatrosses to zebra spiders, they are combined in a common storage cavity at the end of the digestive tract called the cloaca (from the Latin *cluere*, to purge or drain, see page 23), before being released together. This accounts for the multicoloured and multitextured splash of the bird dropping, and why, for example, guano (long-term accumulations of bird excrement) is so rich in nitrogen and so highly valued as agricultural fertiliser.

WHAT GOES IN ONE END...

Dung, droppings, faeces, excrement, whatever you decide to call it (there's a long list in chapter 13), varies considerably from species to species across the animal kingdom, and its final appearance is completely dictated by the animal's foodstuff, and the metabolic chemical processes used to cope with extracting nutrients from it. The nature of an animal's dung will also change if its diet changes.

Arguably, human diets have changed more in the last 50 years than in the previous 50,000. Sugar- or at least carbohydrate-rich foods have proliferated, and processed food has become the norm for many in the western developed world. Where previously 'roughage' was seen as the unimportant nutritional dregs of a diet-plan still rooted in the Stone Age, its importance to human health is only recently becoming clear. Although there are not too many scholarly studies on human excrement, there are suggestions that the vast majority of supposedly healthy westerners are actually chronically constipated. Nutritionist John Cummings wrote a seminal paper on constipation (1984), bemoaning the inadequate intake of dietary fibre, and to lighten the mood introduced it with this superb quote:

> *I have known ... the happiness of a whole household to hang daily on the regularity of an old man's bowels. The gates of Cloacina open, the heavens smile and all goes smoothly. 'Master's bowels have not acted today' from the lips of the faithful butler and the house is shrouded in gloom. Goodhart (1902)*

This is not to be confused with the acute, sudden-onset, constipation that sent me crawling to my local hospital recently. The Bristol stool chart, that handy identification guide to stool consistency, is a neat tool for a broad assessment of our abdominal health. A diet high in fibre, it turns out, protects us from heart disease, fatty build-up in the arteries, diabetes and bowel cancer, as well as averting the painful strains of passing the hard pellets of virtually dry faecal material when we become constipated. Drives to increase vegetable and fruit intake have recently focused on getting people to eat their 'five-a-day' portions of these richly fibrous foods, with moves to up this to seven-a-day, or even ten.

There are anecdotal reports of short-term dietary changes having knock-on effects when it comes to defecation, particularly in the details of aroma and texture, but many of these turn out to be the false bravado associated with drinking too much beer and eating too much curry.

There are, however, several food items which can significantly affect faecal output. In late summer the fox droppings of southern England change from smooth coils of oily grey slime to rough crumbly black and red cylinders, as foxes gorge on the plentiful blackberries ripening in the bramble bushes. In a bizarre study of pig nutrition, Edward Farnworth and his colleagues found they could alter the smell of pig dung by feeding the animals ground Jerusalem artichokes mixed in with their usual swill (Farnworth *et al.* 1995). The dung was lighter in colour, more brown and green, but less yellow, and it was judged to be 'sweeter, less sharp and pungent'. This was partly because it had less smell of skatole, a powerfully aromatic chemical that gives dung its offensive odour – just the thing for assuaging the sensibilities of the pig-farm's close neighbours.

Perhaps the greatest switch in manure output comes with weaning. Anyone familiar with cattle or dairy farming will know that for the first few weeks, when the calf is taking only its mother's milk, its dung is yellow and, if anything, even runnier than normal cow manure – little more than slightly fermented yogurt with a dusting of colour from the bile. When I first saw pats of this at my uncle's North Kent farm many years ago, I thought it looked more like pools of the stuff they use to mark double yellow no-parking lines down the edge of the road. But as the animal starts eating its own food, the gradual change to normal dung occurs. It is at this time that the calf must acquire its own bacterial gut flora. Exactly how cows (or humans, for that matter) obtain their private internal colonies of gut bacteria is still being studied. It is not as simple as accidentally eating food contaminated with faeces; that is a recipe for disaster. But faeces-dwelling bacteria, which then become soil-dwelling, are somehow ingested and soon become established. There probably is something in the old adage that you have to eat a peck of dirt before you die. But you do have to be careful.

As with any animal, part of the trick to survival is learning what you can eat, and what you can't. Contrary to popular rural myth, cows do not just graze everything in their path. Strangely, for those bemoaning the loss of the clover-filled hay meadow, cows should not eat lots of fresh clover. The natural fermentation process in the rumen produces gases which the cow normally releases through belching, but there are substances (as yet not completely identified) in legumes – plants such as clovers, lucern or alfalfa – that upset the bacterial digestion processes. As well as gas, the bacterial digestion of these plants also produces a sticky slime, which creates a thick foam of many small bubbles, rather than a large belchable single bubble. Unable to burp up the gaseous foam, the bubbles remain trapped in the cow's rumen, which threatens to bloat into a balloon to the point of tearing or haemorrhaging, with life-threatening consequences.

The various treatments, devised over centuries (some less sensible sounding than others) include burning feathers under the cow's nose, giving it a pint of gin, taking it for a vigorous run, or placing a stick or rope through the animal's mouth to encourage salivation to break down the foam. As a last resort for a prostrate animal unable to move, the swollen belly is stabbed with a trocar, a large hollow dagger that releases the gaseous pressure build-up in the rumen.

Gases continue to be produced throughout the small, then large intestine, and it is some of these, together with volatile chemicals such as skatole, that give excrement its distinctive smell. There are too many baked bean jokes to make this a surprise to anyone. Two of the most obvious gases are hydrogen sulphide (H_2S), the smell of rotten eggs, and methane (CH_4), the same odourless natural gas piped from oil wells to the gas cooker. The one may offend the nose, but the other is a greenhouse gas reckoned to be 70 times more potent, in terms of its potential to raise global temperatures in the next 20 years, than carbon dioxide (CO_2). Worldwide, livestock is calculated to release over 100 million tonnes of methane a year, and these figures are frequently cited when analysts try to explain climate change. Contrary to comic-book jokes, and quite a few mistaken newspaper articles, most of this gas is released through belching, at the mouth end, rather than with the dung at the anal end, but the chemical processes that give rise to it are a direct consequence of the grass forage being eaten in the first place, and the cellulose fermentation mechanisms that have evolved to digest it. Suddenly, what goes into one end of an animal, and what comes out of the other, turns from being crude schoolboy humour to a serious environmental issue.

CHAPTER 2

CLEANLINESS IS NEXT TO FASTIDIOUSNESS – THE HUMAN OBSESSION WITH SEWAGE

I HAVE A personal history in sewage. When my botanist father, from whom I acquired my interest in natural history, bought his first house in south London in 1957, he looked around his local patch for parks and open green spaces to visit. The closest, and most interesting by far, was the South Norwood Sewage Works. Although not generally open to the public, he obviously got permission from Croydon Council, and in 1961 he published a paper in the *London Naturalist* on the 171 plant species he had found there (Jones, 1961). For the benefit of those readers unacquainted with the mechanics of sewage treatment he started his botanical article with a description of the site and (aided by the works' chief chemist) how the various sedimentation and digestion tanks, biological sprinklers, sediment drying beds, sludge-spreading meadows and filtration fields worked.

When he died in 2014 I found half a dozen old books on sewage treatment, water purification and sanitary engineering still on his shelves, as seemingly out of place amongst the serried volumes on insects, snails, county floras, Darwin and rural economy as his collection of Biggles stories and the Harry Potter series. They were all standard sewerage textbooks from the first half of the 20th century, including the one written by a leading chemist of the day Samuel Rideal (1883–1929) in 1900, which my father listed in the references at the end of his South Norwood paper. In the book was a postcard from the editor of the *London Naturalist* commending him on the spate of articles my Dad had recently sent, and bemoaning the fact that his own recent marriage was taking up too much time to find space to write anything.

By the time my Dad was traipsing through the dense green stands of goose-foot, fat-hen, stinging nettles and tomato plants that dominated the South Norwood sludge meadows, sewage had been treated on the site for 90 years, the scientific culmination of a century of major change in the management of human effluent.

That it needed to be managed, anyone who has ever had a blocked drain will understand instinctively. It is no surprise that human excrement smells, and by varying degrees it is at best unpleasant, at worst downright disgusting. Our visceral disgust (making one's gorge rise, a personal favourite expression) at our own biological output is nothing to do with modern hygienic sensibilities, or Victorian prudery even, it is a far more ancient trait – one that has evolved to protect us from disease.

DON'T TOUCH THAT!

In the far deeps of prehistoric human ancestry, there was no hygiene code to keep cooked and raw meats separate, no understanding of the processes by which provisions go off, no concept of hand-washing to prevent germs being spread from faeces to food, but the look and smell of potentially harmful bacterial breeding grounds left its mark in our revulsion. Humans are naturally repelled by the smell of faeces (our own at least), just as we are by rancid food, and the evolutionary explanation is that this core disgust (which can be visualised in the human brain by MRI scans) prevents the ingestion of noxious materials. Bluntly, any proto-humans who lacked this dung-associated disgust would soon succumb to food-poisoning, cholera, typhoid, polio or a bucketload of other unpleasant and frequently fatal diseases which would, in turn, prevent them passing on their naïve and incautious behaviour to any offspring. Death was ever the prime mover in natural selection. On the other hand, our antediluvian ancestors who avoided their own excrement, because of some instinctive repulsion, survived, flourished and passed on this successful aversion behaviour to their offspring, and eventually down to us.

Similar, but more subtle disgust can also be elicited by seeing sick people, vomit, corpses, bodily sores and wounds, pus, blood and gore, or creepy-crawlies – all potential health dangers for the unwary (Curtis *et al.* 2004). Disgust is a genuine physiological response in the human body, characterized by lowered blood pressure, increased sweating, nausea (sometimes gagging), stopping movement, involuntary shuddering, hair-raising and goose-bumps, exclamations of 'yeuch!' and grimacing. There is no doubt that there is an underlying innate biological disgust response in humans, but this is honed and developed by learning. This is just as well, because it's no good having a disgust that cannot be overridden (unlearned); at least metaphorically, managing sewage is a hands-on activity.

The smell of faeces needs no description, we are all intimately familiar with our own and perhaps our children's. The smell, though, is not from minute airborne particles of the dung itself, but comes from the gases and other volatile substances produced during digestion. As mentioned in chapter 1, the two most obvious gases are methane and hydrogen sulphide. Methane (CH_4) is the most basic hydrocarbon, and the principal component of the natural gas that comes through the cooker hob. Hydrogen sulphide (H_2S) is the smell of rotten eggs, and the usual offending substance produced in schoolboy prank stink-bombs. It is strong smelling and can be stomach turning, perhaps an echo from a time when eating raw eggs could be unhealthy to the foraging *Australopithecus*, if the hatch-by date of potential food items had been passed.

Both of these simple breakdown molecules are produced by the bacterial digestion of food, and although not part of normal human (or other vertebrate) metabolism, bacterial ancestry is such that their microbial progenitors were all busily multiplying and evolving when these two substances were (along with ammonia) major constituents of the Earth's atmosphere a billion years ago and more. The gases are generously spiced with nasally challenging volatiles such as skatole and indole. These polycyclic (6-carbon and 5-carbon rings fused) organic compounds are also produced by bacteria, and from the breakdown of the amino acid tryptophan (one of the 22 essential amino acids that humans need in their diet to build proteins), and they stink. Other equally offensive strong-smelling scents (usually reminiscent of rancid meat) are produced by methanethiol (CH_3SH), and various methylsulphides – $(CH_3)_2S$, $(CH_3)_2S_3$ and $(CH_3)_2S_2$ – from the breakdown of cysteine, a common sulphur-containing amino acid also needed to build proteins. These volatile substances are immediate danger signals produced by carrion.

Humans are no longer so reliant as other animals on their sense of smell, but we will turn our nose up at anything giving off these compounds. What started off as an instinctive evolved avoidance eventually became a much-debated philosophical enquiry into the nature of disease. The germ theory of disease transmission by which invisible microscopic organisms breed and multiply inside a human victim, then get transferred around in excrement, bodily fluids or sneezed droplets, only arrived in human understanding 150 years ago. But for several millennia before this, a vague association between disease and bad smells has bubbled through the history of medicine. Miasmas, lifting off the fetid lands of swamps, sewers, mudflats or any place heavily hung with the scents of decay, were blamed for all manner of epidemic and plague. Malaria, though transmitted by mosquitoes, was closely associated with the marshy swamps around cities, where the mosquito larvae fed in the sewage-rich effluent of the slow-moving waters – the very name comes from the Italian *mala-aria* 'bad airs'.

Whatever the misunderstandings about disease, through all of recorded history the disposal of ordure, and distancing ourselves from its smell, has occupied our thoughts and, to some extent, our literature. In a sparsely populated Iron Age world, the easiest thing would be to take an iron spade, dig a hole and bury the waste. So Moses bade the Israelites:

> *Thou shalt have a place also without the camp, whither thou shalt go abroad; And thou shalt have a paddle upon thy weapon; and it shall be, when thou wilt ease thyself abroad, thou shalt dig therewith*

> *and shalt turn back and cover that which cometh from thee.*
> *(Deuteronomy 24:13,14)*

For the next three and a half millennia this personal self-interring hygiene, followed by the latrine and the earth closet, would be standard practice. When my intrepid traveller friend Mark De Pienne set off on a Namibian safari a few years ago he was similarly instructed to take himself out of the camp with his folding shovel, to bury his deposit when the urge arose one evening. And I'm sure he would have achieved it had not his efforts been thwarted by the appearance of a large menacing snake at an inopportune moment. As he hightailed it (perhaps lowtailed it, I'm not sure) back to the tents, leaving an unfortunate trail behind him, he had other things on his mind than worrying whether his actions would offend the elders, or bring damnation, disease or large predators down upon the trip. When he later recounted his antics Mark never did tell me whether he went back to finish the burial task. But if he didn't, it wouldn't have mattered a great deal, one lost stool in the vastness of Africa.

But away from nomad Africa the fertile crescent of the Near East produced agriculture, farming, settled civilisation, villages, towns and eventually cities. It was the concentration of the masses

Fig. 5 Squat by a wall. The smoke from the lime kiln, and from his pipe, clearly shows the direction of the wind, and the obvious reason why she is holding her nose. From Bewick (1790).

of squirming humanity into huddled proximity, and the burgeoning output of bodily waste, which produced an incentive for the first major advance in sewage processing; it was all about flushing it away, and it has stayed with us ever since.

NO MERE FLUSH IN THE PAN

The fifth of Hercules's twelve labours was to clear out in a single day a 30-year accumulation of manure produced by 1,000 divinely healthy and immortal cattle in the stables of King Augeas. No latrine digging here: Hercules simply diverted the course of the rivers Alpheus and Peneus to wash away the filth. You can almost hear Peisander, the ultimate compiler of the epic poem written about 600 BC, smirking, as he uses the plot device of the flushing water system, just as today's novel writers might invoke a bit of cutting-edge modern technology to spice up a story. It's difficult to know exactly when sewage piping, guttering or drainage channels were first used, but there are archaeological remains from at least 5,000 years ago.

Amongst the oldest claimed toilets in the world are the niches in the Orkney settlement of Skara Brae, inhabited around 3,000 BC. What may be primitive drains were found under the large cells set into the monumentally thick stone walls of the village houses. In the Indus Valley, brick toilets with wooden seats and a chute taking ordure into street drains or cesspits were available for the removal of the effluent of the affluent citizens around 2,600 BC. And by the 18th century BC water-flushed sewage systems attached to toilets are known from Minoan Crete, Pharaonic Egypt and elsewhere in the Middle East.

By the time of the Romans, toilet technology was well established, and the control of flowing water into and out of cities by aqueducts, canals and sluices was common throughout the civilised world. Bath-houses with semi-communal sitting toilets are known from all over the empire, and the Cloaca Maxima ('great sewer') through Rome was a marvel of ancient engineering. It may have started as a series of drainage ditches and an ancient Etruscan canal, but as the city developed from around 600 BC it was soon covered over to create a large arched tunnel under the Forum. It still drains rainwater and debris into the River Tiber in the centre of Rome today. Incidentally, the word 'sewer' comes, via medieval Latin *seware*, from the Roman *exaquare* (*ex* plus *aqua* = water), meaning a place where water is drained off. Cloaca (still used in the anatomical drainage sense in birds and reptiles) comes via the similar-sounding and possibly similarly derived *cluere*, Latin for cleanse.

The disposal of human sewage into drainage ditches and ultimately into rivers and other waterways still goes on, and on the face of it this is a sensible, necessary and environmentally acceptable management process – so long as it does not create pollution. This fine line, between civil engineers' need to get rid of the stuff, and the local residents' desire for clean water, has shifted many times over the millennia.

The Great Stink of July and August 1858 is usually quoted as a revolutionary event in the story of Victorian sewage treatment. The hot weather that summer exacerbated the smell of human faeces and industrial effluent baking and fermenting on the mudbanks of the Thames in London; it interrupted the workings of Parliament at Westminster and raised the spectre of miasmatic diseases striking down the populace of the capital. But this was a scenario that must have been acted out endlessly since the first latrine drained into the stream from which drinking water was drawn.

It's all to do with diluting volumes of water. Modern raw sewage is 99.9% clean water, but this still means vast quantities of obnoxious solids and ammoniacal liquids in the flow. The Thames, though a relatively large river, had a slow flow of water down the shallow geological incline of its broad valley so sewage was only gently wafted on the current. And since it was a tidal waterway, the sewage was then pushed back upriver twice a day, with each tidal surge. The sewage was being flushed away from the houses, but not fast enough, nor far enough.

This was certainly not a modern phenomenon: around 1358 King Edward III was so revolted by the abominable fumes emanating from dung and filth accumulated on the banks of the Thames that he issued a royal charter demanding that no rubbish, at least, should be dumped into the river or its tributaries, and that it should all be carted out beyond the city walls. Semi-liquid excrement would, though, be sloshed or drained into London's streams and ditches for centuries to come, and although many of these waterways, notably the Tyburn and Fleet, were eventually covered over, indeed built over to the point where their original course was lost, they continued to dump raw sewage into the main river until that fateful assault on the capital's nasal senses in 1858. The inner London district Shoreditch is probably a corruption of Soersditch (sewer ditch), which likely drained into the Walbrook; this small river originally flowed through the centre of the Roman walled city, down to the Thames, but is also now covered over and lost.

After the Industrial Revolution, and the increasing movement of people from rural life into the cities, even the faster-flowing rivers of the West Riding of Yorkshire could not cope, and they became smelly and lifeless, little more than open sewers, with filthy water swilling over rancid deposits. In 1866 Charles Clay, an agricultural implement manufacturer from Wakefield, wrote to the Rivers Pollution Commissioners to complain about the miserable plight of the River Calder thereabouts; instead of ink, he wrote his testimonial by dipping his pen into the river water immediately below the town's outflow sewer.

Wakefield is the lowest town on the Calder, which, by the time it arrived there, had also had a chance to receive poisonous outflows from Halifax (via the River Hebble), Huddersfield (via the Colne), Kirklees, Dewsbury and Hebden Bridge, either in the form of human sewage or from the heavy mill industries which used the water as a raw material, or for power.

Problems like this were found throughout the industrialising world. At around this same time Chicago was a small Great Lakes town of about 4,000 people. Its drinking water was extracted from Lake Michigan and its sewage dumped into the Chicago River, which flowed, somewhat unfortunately, back into Lake Michigan. The immense size of the lake meant that there was no problem for many years, but eventually the sewage outflow created a slick large enough to impinge on the drinking water source intakes. The intake tunnels were lengthened to reach into deeper, more distant water, but such was the burgeoning population of the late 19th-century city that its increasing sewage output eventually swamped this measure too. Chicago's ingenious, but environmentally quite shocking, solution was to reverse the flow of the river. By cutting a canal and lock series through to the Des Plaines River to the west, water now flows up the Chicago River from the lake, rather than down into it, eventually taking the city's waste into the distant Mississippi River, a gift for US citizens living on the other side of the great watershed.

JUST ADD WATER?

Disposal by dilution was, until the arrival of industrial cities on the planet, relatively easy to achieve, either by draining sewage into large rivers, or into the sea. Despite some offended sensibilities, it was clear that discharge into water worked as a viable disposal, not just because the waste could be diluted to the point of insignificance, but because there was also a natural breakdown of any organic matter by fish, invertebrates and micro-organisms living in the water, which effectively neutralised any danger from the effluent.

But when the links between disease and excrement were firmly established, and burgeoning human populations became increasingly thirsty for clean piped drinking water, calculations had to be done to work out exactly how much sewage could be diluted into how much water before problems arose.

At first no one could agree. In one celebrated spat, Dr Charles Meymott Tidy, sanitary chemist and barrister, contended that 5% sewage could be purified by natural oxidation in the course of 10–12 miles down a brisk river over a gravelly bed. On the other hand Sir Edward Frankland, professor of chemistry and Fellow of the Royal Society, maintained that a stretch of 200 miles would not do it. Frankland's views were later upheld by the Rivers Pollution Commissioners (of whom he was one), who agreed that there was no river in the United Kingdom long enough to eradicate sewage, even if it were put into it at the source. There was only one answer: sewage from towns or cities had to be removed, treated, broken down or somehow cleaned up. If the final run-off from this process were to find its way back into the rivers and streams from which drinking water was extracted, it was necessary for humans to remove most of what they had put into it.

This, then, is the basic principle of sewage treatment – not just trying to disguise it by diluting it with vast amounts of water, but trying to remove the offending matter. Luckily, help was on hand.

The late 19th and early 20th centuries were a time of unprecedented scientific advancement and understanding of the physical and natural world. Great engineering works were being constructed all over the globe – monumental buildings, intercontinental ships, railways, tunnels, bridges and canals – cathedrals to industry all. Bacteria and other micro-organisms were identified as the culprits responsible for so many human diseases, putting the medieval notions of miasma and bad airs firmly into the museum display case of history, and paving the way for an understanding of hygiene, antiseptics, vaccinations and eventually antibiotics. The very atoms of the universe were being named, measured and ordered, and the quaint alchemical names for compounds such as 'spirits of hartshorn' and 'fire damp' were being displaced by the modern names we still know today (ammonia and methane, respectively), as their compositions were being derived, chemical reactions were being calculated and understood, and new molecules synthesised. All of these fields of human endeavour would be harnessed to get rid of the Great Stink, and smaller stinks across the world.

To know how to treat human excrement in sewage, the first lesson is to hark back to vaguely imagined beginnings, when all humans did the equivalent of Moses' bidding and covered that which commeth from them. The lesson here is that human dung, like animal dung, as we'll see later, is naturally broken down and reabsorbed by the environment, if given a chance. Soil is a far from neutral passive substrate, it is a dynamic biological system full of invertebrates and microbes, constantly breaking down and recycling organic matter from dead leaves, dead animals and animal waste. Below this the subsoil filters out particles and minute changes at the chemical level occur. Finally the bedrock is either porous like limestone, allowing the water to trickle through down to the aquifer from where fresh sparkling mineral water can be pumped or drawn, or it is impervious so that the water flows along the geological strata until emerging clean and fresh at some spring or river source. What sewage management actually does is to mimic this natural process, but on an enclosed and industrial scale.

WHAT GOES IN MUST COME OUT

There are variations, but the main theme is fairly straightforward. First waste water from houses and industry, and rain run-off from gutters and road surfaces, is collected in a series of increasingly larger pipes, drains, sewers and underground canals, where it is directed, by gravity, to the treatment plant. To get it further away from a city than simple downhill flowing will allow, the sewage can be pumped up to a higher level and taken on its way by another series of gently sloping pipes and conduits. The elegant architecture of Victorian pumping stations is still evident throughout much of urban Britain. Eventually, though, the water needs to stop and be cleaned, and although the design of sewage works may be rather more prosaic than prestigious, these are still obvious and distinct large industrial constructions in the landscape.

Here sewage removal and water flow control becomes sewage water purification. To start, large objects such as pieces of wood and

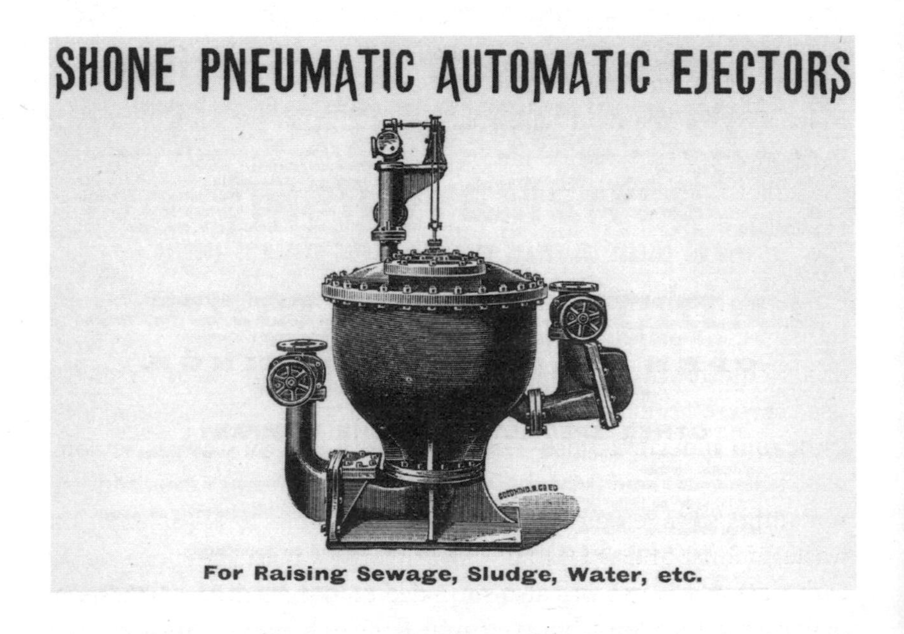

Fig. 6 Every local authority aspired to have a sewage pump, it moved the stuff upwards and onwards. From Rideal (1900).

Fig. 7 When Joseph Rideal (1900) was pontificating about the purification of sewage, the rotary screen for removing large floating objects was the height of modern technology.

other flotsam washed into the drains are caught in a giant sieve. This should also remove paper, nappies, wet-wipes, cotton-buds and any other non-digestibles flushed, rightly or wrongly, down the toilet. Large settlement tanks then allow for heavy solids to sink to the bottom; these include faeces, but also food items sloshed down the sink, together with grit and sand particles washed from roads in storm drainage. Every so often this sludge is dredged or pumped out, drained, dried and spread across fields as a solid manure.

When commercial works like that at South Norwood were opened in the second half of the 19th century, there was usually some aspect of agriculture involved. Fields manured by sewage waste would be used to grow vegetables or other crops, and so the term 'sewage farm' came into widespread use. The process of heavy-duty solids removal was simple, efficient and little more than mechanical, but the remaining sewage water still contained dangerous levels of biological waste product and would fail any drinking water safety test. In the early days this water was also sprinkled onto the soil, across a series of irrigation meadows, but even quite large management plants could be overwhelmed by unexpected flows of water, especially after rains.

Such crops as were grown were unlikely ever to make much of

a commercial return; they were a secondary product in the primary aim of purifying the water, and most 'farms' ran at a loss. Nevertheless cabbages were grown, and fodder plants for animals, including mangold-wurzel, lucerne and grass for grazing and hay. Wheat and potatoes could not cope with the quantities of water being filtered through the soil. Willows, grown for basketry, proved to be too brittle if grown in sewage. Eventually land on the farm would become 'sewage-sick', covered with a layer of slimy sediment and rife with algal growth.

To avoid waterlogging the sewage farm's soil, new techniques for improving and increasing the bacterial breakdown in the water were developed. Fountains sprayed the bacteriologically active water into the air so that more oxygen would get dissolved in the water,

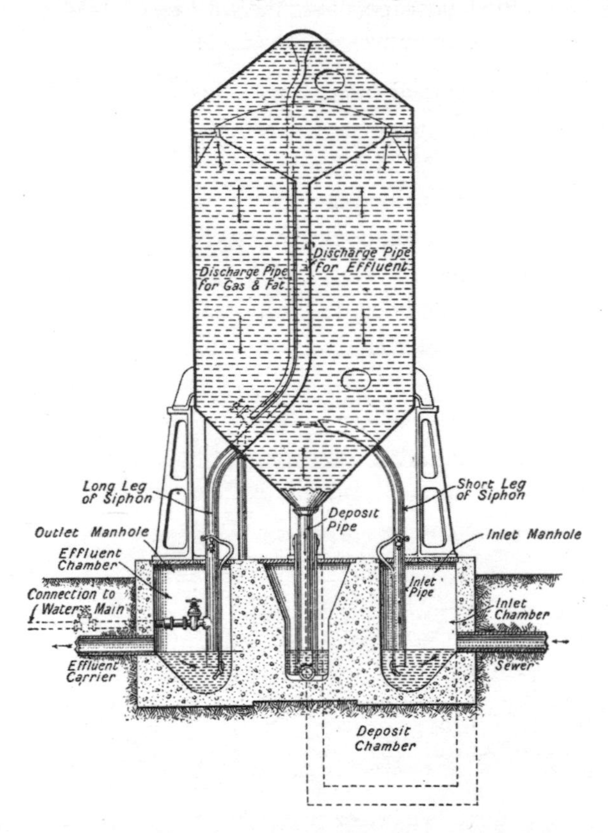

Fig. 8 Kessel separator. Sewage management was the rocket science of the early 20th century. From Martin (1935).

Fig. 10 Fiddian distributors. Circular trickle beds soon became a familiar sight in the landscape, complete with very farm-like houses for the operatives. From Martin (1935).

speeding up the microbial digestion. This water was then drizzled down through beds or towers of sand, gravel and crushed rock to mimic natural filtration. This worked to a point, but filters were liable to become quickly clogged and often needed to be dug up and replaced. My Dad's old textbooks are packed with calculations for flow and filtration rates.

Circular trickle beds became the norm, across much of the UK at least: large broad cylindrical brick-lined or concrete pits are loaded

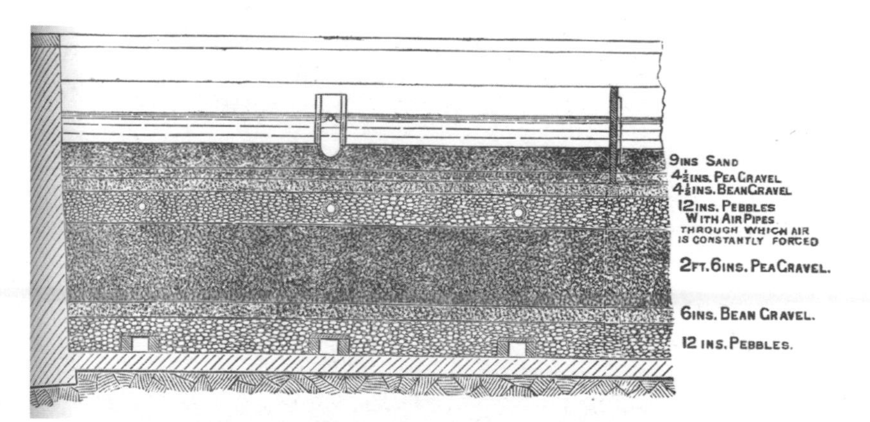

Fig. 9 Lowcock filter. The trickle-down sewage system, worked by mimicking the natural filtration of water as it passed down through the soil. From Barwise (1904).

with small chunks of coke, rock, lava or slag (more recently ceramic, polyurethane foam or plastic) and a rotating gantry above sprinkles the water to percolate down through the medium – not flooding it, but trickling down through the intricate network of surfaces, where air spaces provide access to oxygen throughout the bed. A microbial slime develops on this substrate, its porosity vastly increasing the surface area on which the organisms grow, and it is in this huge three-dimensional well-oxygenated labyrinth that the organic materials in the water are metabolised allowing cleaned water to flow out from beneath.

The cesspit is a much smaller single-dwelling version of this simple technique. The enclosed tank allows sedimentation and as digestion of the organic material progresses, watery material drains slowly through soakaway pipes, or porous brickwork, into the surrounding soil where natural bacterial action continues. Occasionally a build-up of solids needs to be dredged out, but otherwise the limited amount of sewage is gradually biodigested and water returned to the soil.

Cesspits and sewage farms are now relatively old technology, but they suffice, and new sewage treatment centres still work along the same basic principles of solids settlement and removal followed by biological digestion of the remaining tiny waterborne organic

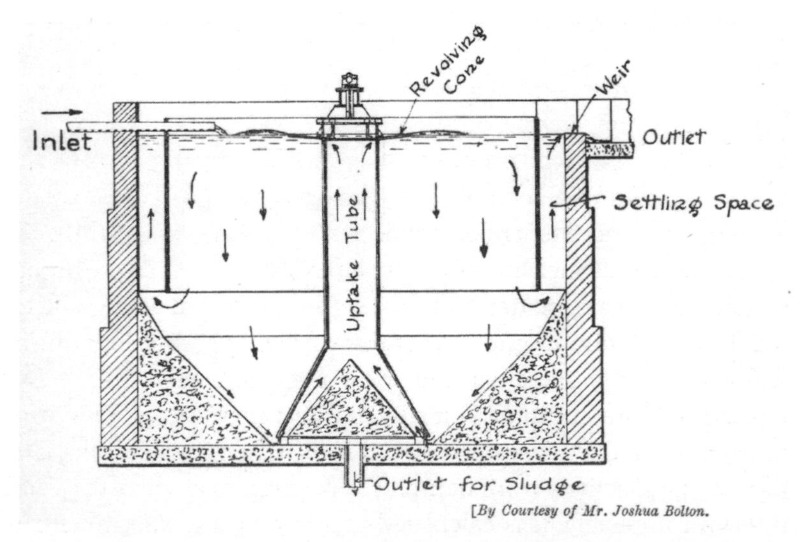

Fig. 11 The modern aeration tank. From Martin (1935).

particles, and the bacteria themselves eaten by other waterborne organisms. In London the water authorities use aeration lanes, large rectangular tanks where air is pumped into the water to encourage the bacterial cleansing; instead of a wet rocky base, the bubbles oxygenate the water, allowing a bacterial soup to carry out the work – a sort of jacuzzi for germs. Sorry. There may be chemical treatments to precipitate specific industrial pollutants, but the business of removing human ordure is pretty down to earth.

In a world where global warming, climate change and altering weather patterns seem set to descend upon us, clean water will become even more of a key environmental issue than it is now. The rain-heavy water-rich northern hemisphere (including the British Isles, renowned for its rainy weather) may soon find itself shaken out of its complacency as water becomes scarcer, and recycled purified sewage water may become the norm for many of us. Drinking our own waste-water may not sound very nice, but how can we even be sure that it is safe?

A TEST FOR PURITY, OR AT LEAST POTABILITY

What we need is a real number, something that can be measured, compared and used to justify that the water being returned to the waterways (or directly into drinking water reservoirs) is clean enough not to cause environmental or medical upset.

In 1912, the Royal Commission on Sewage Disposal came up with the figure which is now the international '20:30 standard'. The numbers refer to a biochemical oxygen demand (BOD) not exceeding 20 mg/litre, and suspended solids not exceeding 30 mg/litre. The suspended solids measure is pretty self-explanatory – it's the amount of fine particulate matter still floating in the water. It doesn't take much imagination to guess what these are particles of, though this will also include silt, dust and other benign materials. The BOD is slightly more complex – it's an indirect measure of how much organic (i.e. mostly faecal) matter is still in the water by calculating the biological activity from naturally occurring bacteria, as they break it down completely. It is calculated in a litre of test water incubated at 20°C for 5 days. If there is little organic matter, it only requires 20

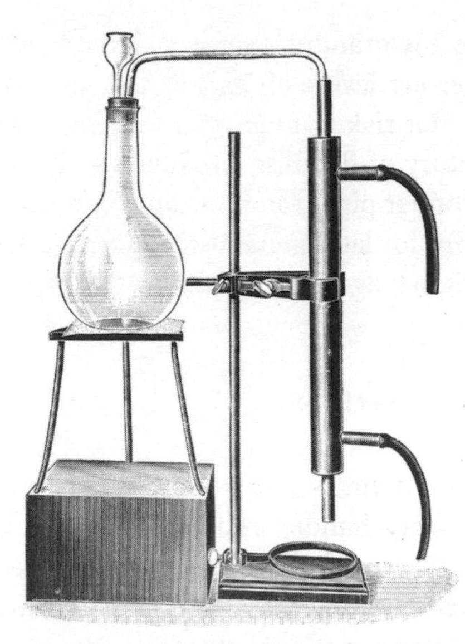

Fig. 12 Sewage treatment was more than just physical removal, it was about measurement and chemical analysis of water before it was allowed to be discharged back into the environment. From Barwise (1904).

mg of oxygen (about a tablespoon) per litre of water for the natural bacteria to harmlessly digest it away. A heavier organic load requires more bacterial breakdown to clean it, hence more oxygen per litre over the test period. Untreated sewage may have a BOD of 600 mg/litre. A pristine clear freshwater river or stream will have a BOD less than 1 mg/litre. The Royal Commission's 20 mg/litre has stood for over a century, and although some people still flinch at the idea of *any* organic material from sewage finding its way back into drinking water, this is still regarded as a 'safe' value.

Back in South Norwood, the purified water from the sewage farm, presumably adhering to or below that all-important 20:30 measure, was eventually discharged into the delightfully named Chaffinch Brook, and then found its way via the River Beck, the Pool River and, the Ravensbourne into the River Thames at Deptford Creek. I've waded in the water of the Creek, and fallen over in it; apart from getting a bit muddy, it never did me any harm.

The sewage farm was decommissioned in 1962, and it was always a surprise to me that it had not been immediately developed for housing; it became the new South Norwood Country Park instead.

I was later told by my grandmother, who lived hard by, that this was because of dangerous levels of lead in the soil, and that the local authorities could not risk home-owners eating vegetables grown in it. Nigh on a century of flooding the fields with water from houses supplied by pre-copper pipe lead plumbing had created its own environmental problem for later generations – a sad irony in view of the sewage farm's original agricultural purpose.

THE YEUCH FACTOR

Today, human excrement is taken away from us at the push of a button or the crank of a handle, and our noses rarely get to complain about being in close proximity to it. But we are not alone in shunning our own bodily waste, or at least viewing it very suspiciously.

Walk through a rich cattle-grazing meadow in summer and it is immediately clear that the grass is not growing evenly. This is nothing to do with the irregular chewing of wayward animals, or an uneven soil layer trampled underneath, or the varied assortment of wild flowers preferentially eaten or distasteful to the tongue, sprouting in the grass. The field seems dotted with slightly taller, slightly greener, slightly lusher tufts of grass, growing seemingly at random across the otherwise evenly cropped sward. Each of these tussocks represents the place where a cow pat fell last autumn, or earlier this spring. There may be some minor increase in grass growth from nutrients being recycled and absorbed through the plant roots, but the major cause of these more prominent tufts is that the cows avoid eating the grass near where their own dung has fallen. Horses act similarly.

The precise reasons and mechanisms of this avoidance remain speculative. There may be a similar disease-avoidance evolutionary mechanism to that at work in humans. Cow dung, especially in wild, or feral breeds, has a heavy parasite load in the form of intestinal worms and flukes, as well as bacteria. Not eating grass near the dung might help avoid reinfection. Oddly, cows *will* eat grass growing from horse droppings, and horses will eat grass growing out of cow pats. Different cow- and horse-specific parasites may mean that ingesting the 'wrong' worm cysts may not present much of a health risk. Cows and horses show no constraint when it comes to eating grass where they have urinated. There is no evolutionary pressure here, since urine does not contain parasites, or indeed bacteria, so there's no biological need to avoid it.

There is nothing wrong with the grass growing out of the dung, and if it is cut and presented to the cattle away from the pat site, they gobble it down. It is the smell of their own dung that averts their noses, and takes their scissor-cutting teeth elsewhere in the field.

Human aversion to dung is not confined to our own waste; we tend to shun all excreta. There are still regular complaints when farmers go muck-spreading, as the aroma of cow slurry goes wafting across the countryside. In the heyday of horse-drawn transport, concerns were raised about the vast quantities of horse dung dropped daily onto the streets of towns and cities. Concerns centred around disease spread and plagues of flies, rather than the smell. Horse dung is one of the least offensive in this regard. George Cheyne (1671–1743), physician, philosopher and mathematician, summed it up neatly in his *Philosophical Principles of Religion* (1715) as he sought to demonstrate the great wisdom of the Creator, who knew that horses would frequently be around humans: '[T]he cleanness, beauty, strength and swiftness of the horse, whose breath, foam [sweat], and ev'n excrements are sweet, and thereby so well fitted for our use and service!'

Perhaps the least offensive dung of any large mammal is that dropped by elephants. These huge animals eat so much herbage, and pass it so quickly that there is some truth in the notion that their excrement is barely more than just processed vegetation. According to trackers, a good way to tell how fresh the dung might be (i.e. how recently the elephant passed this way) is to thrust your hand right into the pile to see how warm it is.

Other dungs we find less or more offensive for some not altogether clear aesthetic reasons. As a general rule, herbivore dung seems relatively easy on the human nose; omnivore dung surprisingly and perhaps unnervingly is very familiar to us (since we too are omnivores), and carnivore dung is especially unpleasant. Anyone who has accidentally traipsed dog dung into the house knows the powerfully evil smell it leaves.

My own personal bugbear is fox dung. When we moved into our present house, my two daughters were aged 3 years and 18 months old, respectively. Whilst we busied ourselves unpacking cardboard boxes of books and crockery, they explored the new garden. The 3-year-old paddled up and down the uneven lawn on a small tricycle, but when she came indoors it very soon transpired that she had stepped in a large semi-liquid ooze of oily grey fox dung out there, and she'd got it all over her shoes. By the time I had cleaned her up I had it all the way up to my elbows. I have been waging war against incontinent vulpines ever since.

Fox dung is especially fragrant, gaggingly so, not because it might spread disease, but to prevent the spread of foxes. It is a powerful scent marker, laid down by the owner of a territory to warn other foxes to keep away. This accounts for foxes' repugnant habit of laying their droppings in obscenely prominent positions – on my front-door mat, at the front gate, on a crisp wrapper dropped in the street, on an upturned paint can left by the farmer at the side of the barn. It is deliberately left in the open, at a prominent spot, so that no one and nothing can miss it.

Other animals make similar statements. Rabbits and hares leave their crottels (also croteys, or crotisings) on ant hills and tree stumps. These small pellets are not at all strong smelling to the human nose, but male and female scents are apparently left in the urine which is also added. Badgers, living in social groups of up to a dozen in a large burrow, combine their massed faeces into a latrine – a series of small oblong holes dug nearby, and constantly replenished and reinforced with their dark, thick, tarry dung (faints, fuants, or archaically werderobe). Again, the purpose seems to be to explain to any wandering badgers that this is a home patch, occupied by a gang of well-fed and well-organised owners, who have staked their claim here.

Otters leave their spraints on small scraped piles of sand or silt at the edges of lakes or rivers, or at prominent places along the stream bank. Despite a diet of fish and other mostly waterside animals, these droppings are not repulsively strong smelling, and although they may be slightly fishy, they also have a sweet musky smell, often reminiscent of violets, or Earl Grey or jasmine tea. Again, their purpose is not to offend or intrigue humans sniffing along the riverbank, but to communicate to other otters in the vicinity subtle messages about fertility, territoriality and body size – the otter equivalent of macho posturing.

CHAPTER 3

WASTE NOT – DUNG AS A HUMAN RESOURCE

HUMANS ARE RESOURCEFUL, adaptable and inventive. Throughout their history, and indeed their prehistory, they have made use of just about every natural product that the planet has to offer. Why should dung have escaped their entrepreneurial gaze? The obvious use for animal dung, and one that is still very familiar to Western readers, is as manure, but before investigating that, it's worth having a look at some of the other less familiar applications, and also some bizarre tabloid-headline niche uses that have been experimented with down the years.

Perhaps the most sensible of the various alternative uses for dung is burning it for fuel. This really only applies to herbivore dung. As soon as the moisture has left it, the neatly chopped and ground herbage particles, processed and conveniently packaged into briquette- or log-sized parcels, have a consistency ranging from balsa wood to beech logs. I know from personal experience that old horse dung and dried rabbit pellets, crumbling and powdery, make good tinder at the camp fire. In North America, buffalo 'chips' were a standard fuel, both for the colonising settlers and for the native first nation peoples. Burned cow pats are still a regular source of domestic heat across much of undeveloped Africa, India and Asia; pats or shaped cakes are dried in the sun, then stacked into architecturally elegant storage piles ready for use.[1] From 1870 until 1976 a steamship, the *Yavari*, plied the Andean Lake Titicaca, its 60-horsepower two-cylinder steam engine famously fuelled by dried llama dung. This is the sort of key fact that tourist guides love to quote, and I can confirm that I am the audience for which such items are intended. If I knew of a dung-powered steam locomotive or river boat nearby, I would have visited by now.

Burned cow dung has also acquired a reputation as an insect repellent, although theoretically any smoke might interfere with the insects' chemosensory detection of blood-laden humans to bite. Alfred Russel Wallace, co-originator, with Charles Darwin, of the theory of evolution by natural selection (and a personal hero of mine), was more than sorely vexed by mosquitoes throughout his worldwide travels; he was frequently incandescent with rage at them, and wrote about their annoying attentions in many of his books. In September 1849, at Para in northern Brazil, he was being troubled, as usual, by their incessant biting when he tried to write up his diary and scientific notes each evening. To find some solace he copied the locals who burned a brazier of dried dung at every door, it being the only thing that kept the midge plague at bay. He noted that: 'In the evening every house and cottage has its pan of burning dung, which gives rather an agreeable odour' (Wallace 1853). It will, perhaps, come as no surprise that experiments to produce insect repellent sprays and oils from cow dung are already underway. It is difficult to know how much credence to give this effort since the dung itself it a powerful attractant of insects, and skatole, the key faecal odour in dung, is also known to be a chemical attractant for some insects including South American orchid bees and various mosquitoes.

In the cool and wet temperate climate of northern Europe, burning dung seems too far-fetched a notion, although there are clear historical records of dung-burning in Brittany up to the beginning of the 20th century. Methane, however, produced by sewage treatment works, and from small-scale manure digesters attached to farms, is increasingly harnessed, stored and burned – a perfectly acceptable modern convenience.

[1] These cakes (called *gobar upla* in Hindi) are widely available on Indian mail-order websites, but for ritual or nostalgic burning by urban dwellers, far removed from the cows of their childhood, rather than domestic heat production.

Another intuitively reasonable use of cow dung is as a building material. Mixed with mud, the chewed plant fibres add strength and elasticity to daub applied to wattle walls, or it can be baked into bricks. As soon as it is dry, cow dung loses its smell, all the pungent gases and volatiles have evaporated away, and it becomes a relatively inert substance. Mixing it with clay gives it a rigid strength, and reduces its flammability. This once purely DIY or cottage industry is being increasingly industrialised, and there is usually dung-derived methane available to fuel the brick kilns, giving an extra environmental cachet to the product.

A once widespread practice that has not continued into the modern age is the use of dog dung in the leather tanning industry. This excremental matter was used right up until the early 20th century, collected from the streets where it fell, usually by children, and was part of the noisome substances, including human faeces and urine (collected in communal pots), combined with the smell of rancid fat and decaying flesh, that made tanneries such evil-smelling places, relegated to the outskirts of cities. The dung was used to soften the leather, through the activity of the bacteria in it. Elsewhere pigeon dung was used.[2]

WHAT WILL THEY THINK OF NEXT?

From these seemingly sensible uses, there now extends a fanciful vista of much odder dung resource possibilities, punctuated with ideas that range from the bizarre to the downright nonsensical. According to tradition, Kumalak, a mystical divination system from Central Asia, relies on a shaman interpreting the patterns exposed when rearranging 41 sheep droppings on a grid of sqaures. Beans or stones can also be used, but since *kumalak* actually means 'sheep dung' in Turkic, I'd feel rather short-changed if it were me having my future read by a proponent using these modern hygienic alternatives.

[2] Incidentally, the oft-quoted eating of dove's dung in the Bible (2 Kings 6:25), costing five pieces of silver for a quarter *cab* (about 300 ml), most likely refers to an as-yet unidentified plant then given that idiomatic name because of some supposed resemblance to pigeon droppings.

Dung, as any macerated and mashed plant fibre, can be readily made into paper. Again, once it is dry it loses any faecal scent, but harvesting sufficient quantities has so far kept this a relatively obscure artisan product. Elephant dung is amongst the most inoffensive of droppings, being little more than pre-chewed plant material, passed in large quantity through a digestive tract that works by removing only small amounts of nutrient, but from a vast throughput. Such was the success of elephant dung paper, that similar products are now created from the dung of cows, horses, moose, donkeys and giant pandas.

On the other hand using elephant dung as a textural splatter in paintings, and for the bulky base of designer platform shoes, is simply relying on novelty to gain popular attention. Likewise moose and deer droppings, dried and varnished and made into drop earrings, also rely on a headline-catching you-cannot-be-serious factor.

The latest use of the elephant gastrointestinal tract is to partly digest coffee beans, which can then be extracted from the droppings and used, either to make drinkable coffee (under the brand name black ivory), or to be brewed into elephant dung coffee beer. There is still a bit of the gimmicky sensationalist venture here, but these are genuine and apparently successful enterprises, used to bring in hard cash and garner international awareness in the struggle that is 21st-century elephant conservation. This is still a cottage industry, but it echoes a similar small-scale operation that is now worth tens of millions of dollars. Black ivory takes its inspiration directly from the natural proclivity of the luwak (the Asian palm civet, *Paradoxurus hermaphroditus*), a small, mostly arboreal cat-like mammal, to eat coffee fruit in its native range across southern and southeast Asia.

The usual history story goes that when the Dutch planters introduced coffee-growing into Sumatra and Java, in the 18th and 19th centuries, they forbade native workers from collecting the fruit from the trees for their own use, such was the high value of the crop. Instead, the local workers collected civet droppings, through which the beans passed almost untouched, cleaned them, roasted and ground them, and made a coffee with an improved aromatic tang and subtle non-bitter taste. It's the protein content of normal coffee beans which gives the drink its bitterness, and although the nut-hard bean looks unaffected during its pass through the luwak digestive tract, the claim is that these bitter proteins are significantly altered by the enzymes in the animal's gut, producing a unique smooth mellow flavour. The fame of the kopi luwak (civet coffee) spread and it soon became a marketable commodity in its own right, selling today for something like £500 ($700) per kilogram; forest-collected wild civet-eaten beans are the highest prized, and achieve the best prices. Civet farming now occurs widely in the region, but disreputably poor animal housing conditions, force-feeding horror stories and fraudulent claims of dubious coffee provenance plague the industry.

In yet another variant, caterpillar teas are available in China – brewed from the frass of caterpillars fed on various plants. This is 'tea' in the broader medicinal sense of a hot infusion of crushed leaves (or in this case partly digested leaves) taken against complaints of the spleen and stomach, to aid digestion, or to ease summer heat, rather than in the popular milk-and-sugar cuppa beverage more typically drunk in the UK. However, some enterprising wag has taken to offering tea-bags of the stuff for the export market. Unfortunately I just missed a recent tasting in the entomology department of London's Natural History Museum, where the general consensus was that it tasted a bit like tea.

I am pretty skeptical about the claims of William Salmon, whose book *The Compleat English Physician* of 1693 gives details of sheep dung tea used to treat smallpox, jaundice and whooping cough. At the time, this extraordinary pharmacopoeia of drugs and concoctions was widely ridiculed for its quackery. However, in a recent *Dictionary of Prince Edward Island English* (Pratt 1988) sheep dung tea is genuinely reported as a folk remedy given to sick people, along with the supporting claim that this is *not* a euphemism. I would happily taste kopi luwak, but even I baulk at the idea of sheep dung infusions. The various adulteration of tea and coffee acts (1724, 1730 and 1776) are widely claimed to have been needed because of assertions that ground-up sheep dung was being added to what was then an expensive high-end commodity.

Many would argue that civet coffee, black ivory and caterpillar tea are just riding on the back of their novelty gimmick value; taste tests are notoriously subjective, and people will try anything new, reassuringly expensive and claimed to be exclusive. Similar claims can be levelled against the idea of facial beauty treatments made from nightingale droppings (developed in Japan) and chicken dung (Hippocrates suggested pigeon) used to treat baldness. All rely on the unreliable feedback of desperate, often gullible customers.

Lion dung is marketed for its supposed ability to prevent deer, rabbits, cats or other unwanted animals from coming onto private property. Intuitively this may seem plausible: the rationale is that these annoying pest animals instinctively recognise the smell of a dangerous predator and stay away. The trouble is that lions are not native through most of the Western world, so no instinctive recognition of their smell could have evolved in the local fauna. According to the website of the British Deer Society, lion dung is very smelly (as is that of most carnivores), but ineffective. To my mind, human excrement would seem a more logical choice to deter undesirables in the garden. It also has the added benefit of attracting interesting dung beetles. Try it.

Urine is another important biological product that humans have utilised over the millennia. Apart from its important role in leather tanning, it was also used as a cleaning agent because after a time the urea in it breaks down to form ammonia; this specially aged urine (lant) was also used for wool scouring (washing). Combined with straw it was an early source of saltpetre for gunpowder manufacture. Alchemists tried to distil gold from urine, possibly something to do with the yellow colour of the liquor, and although this was ultimately unsuccessful, it did lead to the discovery of the element phosphorus by Hennig Brand, in Hamburg, around 1669.

There is a long-running debate, now probably relegated to the status of myth, as to whether Indian yellow, a rare pigment used in oil painting in the 18th and 19th centuries, really was made from the urine of Indian cows fed on a diet of mango leaves. The leaves genuinely contain a substance mangiferin, a glucosyl xanthone, which is converted by herbivore digestion to bright yellow xanthenoids similar to euxanthine, the scientific name of the chemical pigment. However, whether this ever happened on any commercial scale is uncertain, mainly because mango leaves are quite poisonous to cattle, and it would have been just as efficacious to extract the colour by the routine treatments of macerating, boiling with chemical additives and then purifying by evaporation. It is quite likely that the urine anecdote was invented (or at least enhanced) by someone wanting to keep the arcane mystery surrounding a lucrative and exotic commodity.

There is, however, no doubt over the origins and utility of dominant-buck or 'doe-in-oestrus' urines, which are used sold as scent lures for the US deer-hunting market and are widely available in high-street stores there. And in Sudan, cow urine is used as a hair dye, tinting the local Mundari herdsmen's normally jet black hair a scorched red colour. Of course, urine is still widely used, diluted with water, as a liquid fertiliser on the allotment, but those tales about using it on sea-urchin spikes or jellyfish stings are all nonsense.

THROWING IT ALL AWAY

Today we waste millions of tons of our own waste by flushing it away down the toilet. Things were not always so. In the West we now have a tradition of sloshing our excrement off into water, as exemplified by those excellent archaeological sewer remains from Orkney, Crete, Rome and the like, but judging from practices further out across the globe, this is not the only way humans have dealt with faecal disposal.

In 1909 US agronomist Franklin Hiram King toured Japan, Korea and China to examine the permanent agriculture which had been practiced for millennia. In a delightful book *Farmers of Forty Centuries*, published in 1911, he bemoans the wreckless flush-away attitude of the developed world, and enthuses widely about the ingenuity of these great oriental nations. His book is well illustrated with the vessels in which ordure is collected in each house, and how these are garnered by merchants, and shipped by canal barge to agricultural areas for use as manure. King is singularly impressed by the quantities and financial arrangements involved. In 1908 the International Concessions (commercial trade emissaries) in Shanghai sold to a Chinese contractor for $31,000 in gold the privilege of collecting 78,000 tons of human waste and sending it by a flotilla of boats out into the country to be sold to farmers. Meanwhile, in the same year, nearly 24 million tons of human manure were applied to Japanese fields, an average of 1.75 tons per acre. He goes on to extrapolate how many millions of pounds of nitrogen, potassium and phosphorus are squandered annually in the West. His tone is that of an exasperated campaigner, unable to comprehend the profligacy of his native country, and his sentiments would not be out of place in a modern publication decrying the pollution of the waterways and the obvious benefits of self-recycling organic agriculture.

Despite the obvious antiquity of sewage pipes and drainage ditches, sluices and outfall cloacas, such engineering works have only really been viable in towns and cities. Even here, though, erratic piecemeal development often meant that buildings were erected without such utilities. Elsewhere, and especially in rural areas, the latrine gave way to the earth closet, then to the privy attached to a cesspit. Full (or mostly) mains-sewer connection is a relatively recent occurrence.

There have always been a complete mix of sewage disposal options, and one of these has been bucket collection, sometimes for use as manure. 'Night soil', as it was euphemistically termed, was collected at night, and carted off, either for dumping away from the town, or to fertilise agricultural soil by the night soil men, night men, or gong farmers.

That human dung is, or was, used as fertiliser should come as no surprise. Modern farming is still much occupied with using animal manure for this purpose. Every dairy or beef farm, stable, stud or chicken battery will have a dung heap from which material is taken to replenish the soil on which crops or flowers are grown, or animals grazed. At best, this agricultural recycling makes regular use of the various animals' output, but in towns and cities, away from the everyday rural rituals, and populated mainly by people, the emphasis would have been biased towards the getting-rid-of, rather than the making-good-use-of end of the dung disposal spectrum. Very often getting rid of meant dumping in some out-of-the-way place – the midden. The very word itself is a borrowing from Old Norse *myk-dyngja*, literally a muck-heap, helpfully brought into English with the Viking invasions of the 8th–11th centuries, and from which we also get the word 'dung' itself.

Today, the concept of the village or town midden has rather faded from popular imagination, to be replaced by the municipal recycling centre and the landfill site. These modern phenomena have a certain scent associated with them, and usually the sight of wind-inflated plastic carrier bags caught in nearby hedgerows, but in the days of dung-dumping, they must have been truly overpowering places, and any self-respecting metropolitan governing body would want to have them as far removed from the populace as possible.

Of course, the stuff needed to be moved, just as it still was a hundred years ago in the Far East when F.H. King was extolling the sagacity of the oriental ordurers. Read any gritty historical novel (C.J. Sansom, I'm thinking of you here), and with any luck the author will have inserted graphic details of military latrines, public easements, the jakes or other colourful renderings of medieval toilet facilities. This may be augmented by mentions of shit-ditch lane, dung-carts and the heavy stench of human (and animal) excrement as it is taken off from the metropolis to... somewhere. These are based on actual historical documents, but there is precious little trace of such places today. Old castles still show plenty of built-in garderobes, but these were usually just small niches or cubicles with a seat-hole to take effluent straight out into the moat. What must have been quite some considerable industry of manual faeces removal, human and animal, from towns and cities out into the countryside has left small trace on our lives today.

According to most translations of the Bible, Jerusalem had its own dung-port (Nehemiah 2:13) or dung-gate (Nehemiah 3:13, 14, 31) in the city walls. Whether this was where ordure was removed to be dumped out of sight or composted for later use is not clear. In Chaucer's *Nun's Priest's Tale*, the body of a murder victim is hidden in a dung cart, on its way out of the city to manure the fields.

FROM DUNG HEAP TO HILL OF BEANS

Whether produced by human ordure, or farm animal excrement, the dunghill or dung heap was a familiar part of any agricultural landscape in the world. Again, there are plenty of biblical, Chaucerian and Shakespearian references – thank heavens for searchable online texts, eh? Although disguised by recent changes in language, midden, dunghill and manure are frequent origins of place names, on both larger and smaller scales. Maxfield, near Eastbourne in Sussex was Mexefeld in the 12th century, *meox* being old English for dung. Terwick, also in Sussex, was Turdwyk in 1291, and is from Old English *tord* (turd) and *wic* (farm). In 14th-century Bedfordshire, le Shithepes, is pretty self-explanatory, similarly Sithepes in Cambridgeshire. Middyngstede in Yorkshire (1548) and Myddenhall in Dorset (15th century) are just two of many places named after the midden. Closer examination of old documents and maps shows that there was a whole slew of field names, obviously connected to manure heaps and the regular manuring of agricultural land (Cullen and Jones 2012).

Today, manure heaps, dung piles and slurry pits continue to be a managed part of modern animal husbandry. As well as getting the smelly and sloppy contents of the stables and cowsheds out of the way of the housed animals, the mounds of composting material are available to the farmer when it is muck-spreading time in the fields, though human faeces no longer feature heavily in this process in developed countries. Having said this, in my 1930, 11th edition, 45th thousand printing of Primrose McConnell's famous *Agricultural Note-Book* (first published 1883) human 'egesta' is listed in the tables to help calculate the spreading rates, and the nitrogen, phosphorus and potassium contents of various manures, along with night soil, both solid and liquid combined, and dried, also town sewage and dried sludge. Presumably McConnell was addressing the sewage farmers as well as the traditional farmers with his popular publication.

Although almost all types of animal dropping can be used in spreading manures, cow dung is the default option across most agriculture, at least in the UK. Produced in copious quantities in the farmyard and milking pens, its semi-liquid nature (77–85% water) means that it can be eased by gravity, and a person pushing a shovel or broom, into slurry tanks or what is sometimes optimistically called a lagoon. Here it can sit, gently fermenting for a few days, before it is pumped out into tankers, muck-spreaders, sprayers or whatever device the farmer is using to apply it to the fields. As I sit and write this in London, in April 2015, my 10-year-old son is getting ready for a week-long school trip to a farm in Devon. He knows what to expect, having heard it from his two sisters who also went there. Despite the bottle-feeding of cute baby lambs, feeding the pigs and chickens, collecting eggs, milking the cows, and going for bracing country walks, the single most powerful memory all the primary school children come back with from the farm is the orchestrated shovelling of cow poo across the yard – the runniness of the stuff, the high smell of it all, and who slipped over in it. This is the stuff of schoolchild legend.

Pig dung, at 72–75% water is also relatively manoeuvrable from the farm yard or sty, but more powerfully pungent, unless the animals are fed on Jerusalem artichokes perhaps.

Horse manure is the favoured dressing and soil nutrient for allotment gardens and smallholdings, and continues to be offered virtually free to passing gardeners by stables up and down the country. Its relative dryness and inoffensive fruity smell make it easy to cart off in the car in plastic bags. Some years ago, when we lived in Nunhead, one of my neighbours was bemused to see me scoop up for the garden the deposits left by two police horses that had recently passed down the street. Well-rotted horse manure is also a popular substrate choice for mushroom growers.

Battery chicken houses produce tons of chicken droppings, which are collected, pelleted and sold for domestic fertiliser. Because faeces and urine are expelled together in a single dropping via the cloaca, as all birds do, chicken manure contains very high levels of phosphate, nitrate and potassium, key nutrients for plant growth.

Manuring was especially important on 'poor' soils. The chalk downs, limestone hills and rocky uplands of the UK sometimes have only a few centimetres of topsoil, low in nutrients and easily lost to wind and rain when ploughed. Sheep-grazing was the only possibility on some of the steeper slopes (it still is), but this fitted into a neat regime to bolster the fertility of the more gentle slopes where arable crops could sometimes be grown. During the day the sheep were free-roaming on the steeper hillsides, under the watchful eye of the shepherd, but at night they were dog-driven then penned in fallow fields nearby where their droppings would fertilise the soil for the crops grown during the next cycle. Such was the importance of sheep on the chalk hills of southern England, not just for wool and meat, but as mobile manure carriers, that they were similarly day-grazed on rich water-meadow grass in early spring and again folded on the fallow chalk land at night. In the mid-19th century sheep were increasingly given oilcake feed (from the waste residue of crushed oilseeds), not just to improve their mutton output, but because the cake paid for itself in the superior manure that they produced on cereal-growing fields (Bowie 1987).

DUNG WORTH FIGHTING OVER

Today the value of the world manure market can only be guessed at. Figures reaching scores or hundreds of billions of US dollars are regularly bandied about. Such is the international scale of the manure trade that governments have legislated on it and wars have been fought over it. In 1856, the USA passed the Guano Islands Act, by which US citizens were empowered to take possession of any unoccupied guano islands they encountered in their travels, and over which the US would then offer military force should these interests need protecting. The 'Guano War', more precisely the Chincha Islands War (1864–1866) began when Spain seized these guano-rich islands just off the Peruvian coast, much to the displeasure of Peru and Chile. The War of the Pacific (1879–1883) between Bolivia, Chile and Peru was sparked by disputes over boundaries in the Atacama Desert, where significant quantities of guano had been discovered.

Guano, from the indigenous South American Quencha word *wanu* (or *huanu*), comprises accumulated deposits of bird excrement, mainly from the guanay cormorant, *Phalacrocorax bougainvillii*. The extremely dry climate of western South America, in the rain shadow of the Andes, allowed the droppings to amass over millennia, neither decaying, nor being recycled by natural processes, nor being leached by rain, and deposits achieve depths of 50 m. Guano is easily mined and transported, and since it is dry and relatively odour-free it makes an easy and convenient fertilizer high in nitrogen (usually as ammonium salts), phosphorus and potassium. Guano remained a primary source of agricultural fertiliser until the early 20th century; in 1869 over 550,000 tonnes were mined. In Peruvian history, the period 1845–1866 is referred to as the 'Guano Era'. It was only eclipsed when the Haber–Bosch process was able to make ammonia from atmospheric nitrogen on an industrial scale using a high-pressure furnace and beds of metal catalysts. Guano is still mined and marketed as fertiliser today, but on a much smaller scale.

Of course, all of these commercialised and targeted manure and guano possibilities hinge on there being animals contained in barn, sty, shed, coop or stable, or age-long accretions of natural droppings, from which the biological matter can be cleared and piled up, packaged, transported and stored ready for use. Out in the field or meadow or wood where free-range animals roam, the dung lays where it is dropped. The soil will eventually get the benefit of the recycled fertilising nutrients, but first the dung attracts the attention of other users.

THOROUGH ROOM DISINFECTION

BY MEANS OF

DRY FORMALIN TABLETS

as described at the Congress of the Sanitary Institute held at Leeds, 1897, and the British Medical Association, 1898.

ALFORMANT "A"

(Regd. Trade Mark)

This Alformant is now almost universally adopted by the Medical Officers of Health in this Country.

Entirely Superseding Sulphur.

It is also the cheapest and simplest method of Thorough Disinfection of the present day.

SCHERING'S PURE FORMALIN, 40%.

Clean, Effective, Non-Poisonous.

ONE GALLON added to THIRTY-NINE GALLONS of Water makes FORTY GALLONS of a Powerful and Unequalled Disinfectant and Deodorant Fluid.

For Literature and Particulars apply to

THE FORMALIN HYGIENIC COMPANY Ltd.,

9 and 10 ST. MARY-AT-HILL, LONDON, E.C.

ALFORMANT. A.
R.L.P
THE FORMALIN HYGIENIC Co. Ltd.

By Royal Letters Patent.

CHAPTER 4

IT'S WORTH FIGHTING OVER – DUNG AS A VALUABLE ECOLOGICAL RESOURCE

IN NATURE, NOTHING is wasted – not even waste. The throughput from herbivores is remarkably undigested, despite the range of complex digestive enzymes, acids and alkalis, convoluted and copious intestinal tracts, and the employment of countless millions of micro-organisms to break down the tough cellulose and complex plant chemicals. It's all that roughage they eat. Typically only 10–30% of nourishment is taken from food into the average herbivore animal's body; the remains are shunted out as dung. Sometimes nutrient absorption is as low as 5% – they've hardly touched their food. Omnivores, if they are following modern nutritional advice, also pass much that is still available for onward use by others.

Within the dung certain items, coffee beans for instance, or the sweetcorn niblets of schoolboy humour, pass through apparently unchanged. Baboons and birds frequently forage through elephant dung to get at any seeds that, protected by tough outer shells, have survived the digestive process. Throughout the world this type of behaviour is common, if only infrequently commented upon, with birds (sparrow, yellowhammer, chaffinch and linnet for example:

Popp 1988) taking seeds and undigested oats from horse droppings, harvester ants removing fruit seeds from capuchin monkey faeces (Pizo *et al.* 2005), and spiny pocket mice taking seeds from cow and horse dung (Janzen 1986). Some seeds will only germinate if they have been passed through the gastrointestinal tubes of a large herbivore, or granivore bird (or both maybe). If the seeds escape these secondary feeders, they benefit by being scattered away from the potentially overshadowing parent bush, and the seedlings will end up further separated, diluting any obvious competition with each other, and enhancing future genetic diversity. The further seed dispersal abilities of dung beetles will be looked at later.

Seeds aside, most of the content of elephant dung is little more than semi-processed leaf material, cut into convenient short pieces, partly chewed to a soft consistency and certainly only partly digested. A small group of four or five elephants can take a tonne a day of rather inaccessible coarse herbage, tear it down, mash it up and deliver it in neatly arranged parcels at ground level, where it is now available for many others to use. In Central Africa, the sitatunga or marshbuck, a type of deer, forages on living shoots and leaves, but also gets a significant part of its nutrition, seeds and chewed herbage, from elephant dung, appearing out of the jungle as soon as the fresh

Fig. 13 The marshbuck, or sitatunga, considers elephant dung as just a convenient silage drop.

droppings are released. It's getting as much from these deposits as would a cow from the farmer's morning presentation of hay or silage.

Elephant droppings provide a good baseline model for understanding the immense biological importance of dung. That it is important is in no doubt, since huge numbers of dung beetles (and flies) swarm in to any available dropping. Incidentally, in the identification gallery (chapter 12) later on, it is obvious that dung-inhabiting insects come in many shapes and sizes, but for most purposes 'dung beetle' refers to any of the approximately 9,000–10,000 species of short, squat, rounded or cylindrical, powerfully built beetles in the hugely diverse families Scarabaeidae, Geotrupidae and Aphodiidae. It is these beetles which will get the lion's share of attention throughout the rest of this book. I admit that this is partly because I find dung beetles fascinating; they are handsome and charismatic. But also because a natural history of dung can readily be distilled down to a natural history of dung beetles. I will, however, try not to let this bias take over too much. Anyway, back to elephant excreta.

THE MAD SCRAMBLE FOR POSSESSION

There are verified reports of nearly 4,000 dung beetles arriving at a half-litre sample exposed for 15 minutes – that's nearly four and a half beetles per second (Heinrich and Bartholomew 1979b). In a timed study, a 30 litre pile (think large kitchen bin) of elephant dung was reduced, in under half an hour, to a spreading fluidised layer of squirming beetles, covered by a thin rind of fibrous material, the insects reducing the beachball-sized heaps to a nearly circular mat 2 metres across and 2–3 cm deep. Elsewhere 16,000 dung beetles spirited away 1.5 kg of elephant dung in under 2 hours (Anderson and Coe 1974). The literature of dung beetles is littered with these astonishing figures. Dung beetles are keen, very keen indeed.

Such observations are not limited to the tropical savannah. In a pitfall trapping exercise in Langley Wood, Wiltshire, Michael Darby was more than a little amazed to find his sampling pots jam-packed with one of the dor beetles, *Anoplotrupes* (*Geotrupes*) *stercorosus*, every time he emptied them. The dors are northern Europe's largest dung beetles and although widespread you seldom find more than one or two under a cow pat or horse dropping. He was finding hundreds at a time in the small pitfalls, and between August 2006 and July 2007 had counted over 20,000 specimens (Darby, 2014).

I can't quite compete with that, but on the South Downs of Sussex one hot August day back in 1974 I watched a cow pat, just a day or so old, heave as several hundred small mottled dung beetles, *Aphodius contaminatus*, jostled together in the congealing semi-fluid interior under a thin rind of dried dung.

Typically one elephantine bowel evacuation lands in a large heap weighing 10–20 kg. This is a finite amount, and although a massive volume compared to even the largest dung beetles (weighing about 20 g), these insects must act quickly if they want to use it. Latecomers will find that they are just too late. The important ecological imperative here is not that they are going to eat much of it (in fact adult dung beetles eat very little, see page 70), but that they need to secure a large enough portion of it to see any of their offspring through from egg to adulthood.

However, if the beetles simply arrived, mated and laid their eggs willy-nilly, it is quite likely that within a very short time the resulting larvae will have devoured all the available dung, even a massive elephant deposit, long before they are mature enough to pupate and metamorphose into adults. They would all perish. This would not be a very successful evolutionary strategy. Dung-feeders have evolved various means to carve up the dung pie, as it were, and these will be looked at in more detail in the next chapter. What is important is that they actually do get a part of the pie, and that they get it quickly.

FIRST FIND YOUR DUNG – AND BE QUICK ABOUT IT

I recently had a welcome occasion to study British dung beetle eagerness first-hand. The beetles were arriving at a freshly dropped stool (that would be *stercore humano* in ultra-polite parlance) one warm May afternoon in 2015. The site, a former railway sidings and shunting yard, had been partly developed and landscaped with mounds of soil, gravel, crushed brick and concrete, to create an 'ecological area' for native wildlife; it had greened up a bit since the municipal recycling centre had been built, with a thin haze of scant herbage, but was well away from any grazing animals, meadows or indeed green fields in any conventional sense. It was here, in a private corner, that I was able to conduct an experiment in the island biogeography that best exemplifies just how incoming colonisers find a tiny portion of suitable habitat in a sea of barren wilderness. In less than 30 seconds from deposition the first dung beetle had flown in and was hovering over the deposit. At just under 3 mm, *Aphodius pusillus* is one of Britain's smallest dung beetles, and was probably more used to the rabbit pellets amongst the sparse vegetation of the site. But it was not alone.

Within 15 minutes, over 50 beetles, comprising nine different species[1] had arrived by air, along with a loudly buzzing selection of blow flies. This partly begs the question where all these coprophagous insects came from to arrive so quickly at a large omnivore dropping where previously there had been no obvious community of dung-producing animals to supply their needs. None of the beetles was particularly large, but several were in the 6–8 mm range, suggesting that rather than rabbit crottels it was horse and cow dung that had sustained the thriving populations hereabouts, and that they had probably travelled many hundreds of metres since leaving the farmland where they grew up. It seems highly unlikely that they had all been busy at work in a meadow half a kilometre away to the south, when they suddenly caught a whiff of something tasty in the air and decided to investigate; these were beetles already somewhere in the close vicinity, they had been active, on the wing, possibly even cruising on the lookout for new supplies. They had been rewarded, and an hour later the dung was riddled with an eager mass of beetles.

Others have carried out similar experiments, albeit accidentally. When coleopterist Clive Washington had to call in a drain company to unblock his household sewer, removing the manhole cover revealed a logjam mass of putrid brown sewage. Within 10 seconds *Aphodius* were flying down and landing beside it, much to his delight.

On holiday in Costa Rica a few years ago, I took whatever opportunity I could to examine the droppings I came across in the

[1] Someone is going to ask, so for their benefit: *Onthophagus coenobita, O. similis, Aphodius ater, A. equestris, A. erraticus, A. prodromus, A. pusillus, Sphaeridium scarabaeoides.* Also the predatory rove beetle *Ontholestes murinus.*

tropical rainforests (I didn't add any of my own on that occasion). There weren't that many – a few meagre nuggets from deer, monkeys and peccaries. Many had a selection of dung beetles hollowing them out from the underside. These were mostly small, squat, smooth, round, slightly metallic scarabaeids. I also found the same beetles sitting on leaves in the shafts of dappled light reaching down from the canopy to the forest floor below, where they seemed to be perching, antennae outstretched all aquiver – waiting. *Canthon viridis* is well known for being very keen, and this perching behaviour is frequently reported. It is, figuratively at least, on its marks ready, set to go, as soon as it detects the chemical scents from a dropping being freshly dropped.

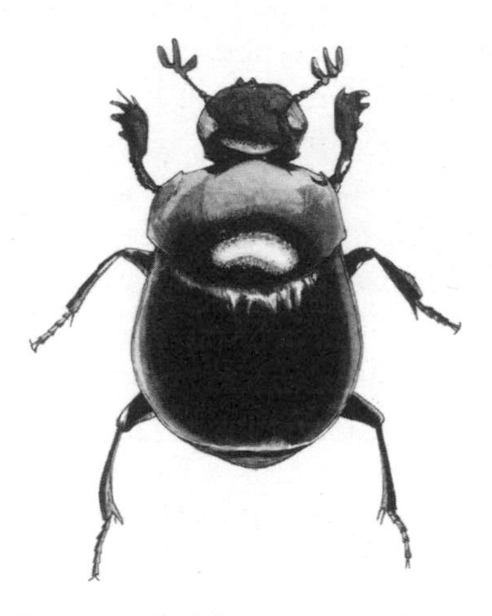

Fig. 14 The splayed antennal club is a sure giveaway that the beetle is detecting faecal scents in the air.

That dung beetles smell out the dung has long been known. When they arrive at a fresh pat, they invariably fly upwind, following the trail of scent that the breeze is bringing down to them. This is an observation which can be repeated by any keen scatologist reader in grazing meadows all over the world. Larger beetles can be seen zigzagging in to the target; they are programmed to fly forwards if they can smell their quarry, but move left or right if they lose the scent for a moment, until they find themselves back in the airborne plume. To visualise this process imagine that you're seeking a smoky bonfire, but that you are blindfolded. If you can smell the smoke you move up into the wind that brought the soot into your nose. But if you step outside of the irregular wafting smoke stream, your best bet of finding it again is to move increasingly right and left, scanning the scentscape perpendicular to your previous route, until you step back into the cloud. Then you set off windward again. This is exactly what scenting insects do when flying towards their quarry.

Dung beetles don't have noses; their antennae are olfactory organs. The last few segments (usually 3–5 of them) of the antennae are flattened and expanded into a series of plate-like slices (lamellae) forming a broad club. This greatly increases their surface area, and thus also the number of submicroscopic chemo-sensitive receptor cells which cover them. These highly sensitive pit- or hair-like structures allow the beetles to smell airborne molecules at incredibly low concentrations – we're talking parts per billion here. Males and females have similarly shaped antennae, suggesting that it is food they are smelling out. In other insects large, highly sensitive, feathery antennae are usually associated with only one gender, which needs the extra smelling capacity, afforded by increased surface area, to detect sex scent pheromones released by prospective mates. Some dung beetles do use pheromones to communicate between the sexes, but finding each other happens at the dung source, so their most important olfactory test is to find the dung in the first place.

Recent electroantennographic research, using isolated antennae in micro-windtunnels, has allowed direct electrical measurement of nerve impulses as different test chemicals are wafted past. Dung beetle scent receptors responded to a range of dung volatiles, including our foetid friend skatole, and other similar bacterial decay molecules such as 2-butanone, phenol, *p*-cresol and indole. Butanone seems to be one of the most important of these molecules; it's certainly the most volatile (boiling point 79.64°C) and has the simplest molecule $CH_3C(O)CH_2CH_3$ (methyl-ethyl-ketone), whereas the others have more complex ring structures. According to some researchers, isolated and concentrated butanone smells of butter-scotch and nail varnish remover (Tribe and Burger 2011).

It seems likely that dung beetles are not simply attracted to just one chemical, but their antennae are sensitive to the melange of different odorants given off by decay. This is useful because it means that dung beetles can be attracted for sampling purposes by the simple expedient of plopping some excrement on the ground. But the fact that each type of chemoreceptor on their antennae reacts to just one type of airborne molecule does, however, throw up some unusual observations. Michael Darby's 20,000 dor beetles were not attracted to dung; his traps were unbaited, but were laced with a small amount of ethylene glycol. This is standard car antifreeze, and as well as being cheap and easily obtained, it makes a convenient preservative in pitfall traps. What may have been happening is that the dor beetles were attracted to the corpses of other insects fallen into the traps (ground beetles especially), which were releasing similar, but non-dung organic decay molecules. Who knows, maybe 2-butanone was among them. This created more corpses, which escalated the effect, leading to the huge numbers of dors eventually accumulated. Elsewhere among dung beetles there are strange examples of species being attracted to chemicals that are not given off by dung, and this has led to dung beetle evolution away from dung to use other food sources. More on non-dung beetles later.

Though they smell the dung from way off, dung beetles use eyesight to target the dropping when they get close. There is a neat relationship between small beetles (with small eyes) which fly by day and can land accurately in daylight, but large beetles (with correspondingly bigger eyes) having the greater optical capability necessary to fly at night. Day fliers are fast and sure, but night flights offer the problem of hazy obstacles, difficult to make out in the gloaming. Slow fliers might open themselves up to predators such as owls and bats, and sure enough nocturnal dung beetles tend to fly fast but sloppy.

Dung beetle eyes are protected by a bar across the middle, on a line with the broad shovel-like plate that defines the front of the head. Called the canthus, it sometimes nearly divides the eye into two, and is thought to offer some defence against abrasion to the eye when digging through the soil, a bit like bull-bars on off-road vehicles might protect the headlights against knocks. Nocturnal species, needing all the visual help they can get, have the canthus smaller, or absent, so more eye is available to catch the fading rays.

NOT PUTTING ALL YOUR EGGS IN ONE BASKET… OF DUNG

These observations of manic arrival at the pat typify the struggles which all dung-feeding insects face – that dung is a rich organic food, but it is produced sporadically, in a random, widely spaced scatter across (to an insect) a truly vast landscape, and that the small dropping from anything other than an elephant is barely enough for a few individuals to raise just one generation of offspring. In ecological jargon, dung pats are regarded as patchy and ephemeral microhabitats. In order to find them before they have decayed, or dried out, dung-seekers need to be active, adventurous and willing to take to dangerous skies full of wayward winds and hungry predators, expending energy and possibly risking their lives, if they are to find a competitor-free food source for the next generation.

It's slightly cheating, but some dung beetles don't need to find the dung themselves, they are brought to it by the dung depositors. It is immensely sad that the common name 'sloth anus beetles' has not been universally adopted for the genera *Uroxys* and *Pedaridium*

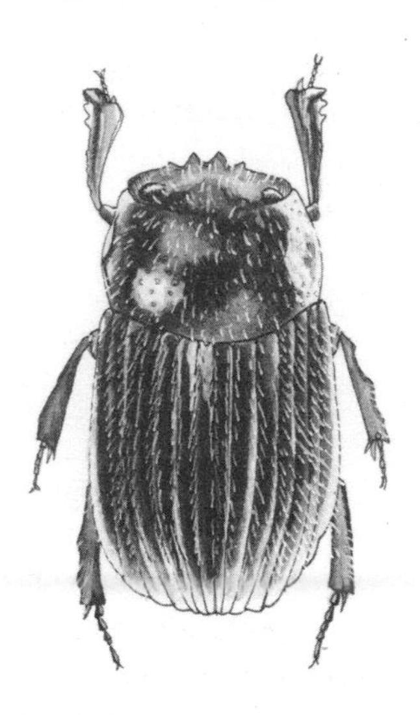

Fig. 15 It took entomologists some time to work out how the tiny (3 mm) *Pedaridium* sloth beetles lived. They travel in the coarse sloth fur, but hop off to lay eggs each time the host descends from the trees to deposit some dung on the ground.

(formerly *Trichillum*) which, as their name might suggest, hang around on hairs near sloth anuses, waiting until sloth defecation takes place (Ratcliffe 1980; Young, 1981b). Similar beetles in the genera *Glaphyrocanthon*, *Canthidium* and *Canthon* loiter near monkey anuses, and several *Onthophagus* species cling to the nether orifices of wallabies and kangaroos. They have specially adapted prehensile claws to stick tight.[2]

However they arrive, dung beetles will have to compete against similar-minded contenders for the resource. Even if they succeed in getting there and laying eggs, their offspring will have to struggle against the demands of all the other dung inhabitants, including each other. Too many of them, and they will, quite literally, eat themselves out of house and home. They need to find the dung and get in quickly, they need to stake a claim, and at the same time they need to ensure that they will not overburden the very limited food resource they have been offered.

It turns out that although dung beetles are highly mobile (apart from the anal hangers-on), energetically active and quick to home-in on fresh faeces, they are not overproductive in the egg department, and that they do not, as it were, breed like flies. Actually, flies also have similar pressing needs when it comes to securing a dung-sufficient future for their maggots. We'll come back to flies a bit later.

Of the few species that have been studied closely, an individual female dung beetle is highly unlikely to lay more than a few dozen eggs during her lifetime of many months. And although some of the smaller general dung-feeders may eventually manage 150 eggs, many of the large exotic dung-rollers that rush in to an elephant dropping are only ever likely to produce 5–20 offspring each, often only one at a time. The nesting behaviour, examined more closely in the next chapter, where individual buried balls of dung are attended by a guardian female, appears to have an egg-suppressing physiological effect on the beetle herself, delaying any further eggs in her ovaries from becoming mature until she is convinced that she has done all she can there and is ready to move on to start again. Already there are

[2] In a similar extreme association, the minute specimens of *Acuminiseta pallidicornis* (lesser dung fly family Sphaeroceridae) ride around on the backs of giant African millipedes in the jungles of West Cameroon, waiting for their hosts to deposit large (to the fly), invitingly moist frass droppings (Disney, 1974).

self-limiting restraint mechanisms at work here, to prevent single dung pats being overwhelmed, to the detriment of all.

WHAT IS THE POINT OF HORNS?

A self-evident indication that dung is a valuable natural resource is the fact that dung-feeders not only swarm up to it as soon as it lands on the ground, but that they are apt to fight over it when they get there. The petty squabblings of dung flies will be looked at again shortly, but some dung beetles have armed themselves with quite formidable weapons in the long evolutionary battle for niche supremacy.

The ridges, humps, bumps, spikes, spines, horns, prongs and antlers borne by many male dung beetles are some of the most extravagant body decorations in the insect kingdom. They range from the stout but elegant single head horn, through bulbous spikes and ridged skewers, to flanged and fluted art nouveau protuberances, and they're well worth looking at if you need inspiration for an unusually gothic Hallowe'en costume. The fact that it is generally males that carry these weapons, and that they are attached at the front end – head and thorax – clearly supports the idea that they are used in fights. They're analogous to the impressive antlers sported by male deer, though unlike testosterone-pumped bucks during the noisy and dangerous rut, dung beetle head-to-head attacks are rarely witnessed.

There were some suggestions that these over-exuberant horns are the result of sexual selection – choosy females choosing only to mate with the best endowed males. Charles Darwin first mooted this idea in 1871, imagining that these horns became ornately developed through natural selection in the same way that increasingly bigger and brighter peacock tail feathers were chosen by an ongoing selection of particularly picky peahens down the millions of generations. However, morphological and behavioural studies now suggest that dung beetles' wondrous projections are genuinely used for combat. These are jousting and wrestling contests, pushing and straining, rather than the insect equivalent of sword fights, and the contests take place between males of the same species.

It's worthwhile spending a little time examining dung beetle horns, because they are truly fascinating. They're beautiful, weird and there are some very interesting ecological lessons to be learnt from their study.

Different dung beetle species will have different approaches to using the dung, different techniques in extracting it, different needs in terms of size, density and moisture content of the dung. They may compete, one species against another, but such scores are usually settled by subtle ecological and behavioural differences – they arrive at different times of day, or in different seasons, they may live in different places, or take different portions of the pat. On the other hand, beetles of the same species will have directly competing needs, and it is this within-species conflict that has given rise to the dung beetle tusk-based arms race. Simmons and Ridsdill-Smith (2011b) and Knell (2011) give good overviews of horn evolution and development.

The very large genus *Onthophagus*, with upwards of 2,500 species worldwide, provides plenty of good examples, with something like 10 different standard shapes of horn or antler, arising from 25 different zones of the anterior body surface of the beetle. These different sizes, and combinations of head and thorax horn, create an almost infinite number of permutations, much to the delight or consternation of the taxonomists trying to identify these beetles. Many a time I've pored over the microscope, trying to convince myself, or not, that the subtly different spike shapes on a series of *Onthophagus coenobita* males represented different species, or different subspecies even. Frustration and delight in almost equal measures. Numerous dung beetle researchers brazenly assert that *Onthophagus* is the most species-rich genus anywhere in the animal kingdom. There certainly are a lot of them, but rival claims are also made by jewel beetle (Buprestidae) specialists for *Agrilus*, arguing 3,000 species. I'm not going to take sides in this discussion, but I suspect arguments over the validity of genera, subgenera and species groups could keep these two clans at loggerheads for some time.

And keeping each other at loggerheads (see what I did there) is just what the beetles do when they fight over the dung. Actually, they fight under the dung. Whatever their weaponry, two dung beetles meeting face to face on a flat laboratory surface fail to engage each other, but in the confines of a burrow, or space under the dung, head-to-head conflicts can now escalate. The beetles can shuffle, one left, one right, or one upside-down compared to the other, and

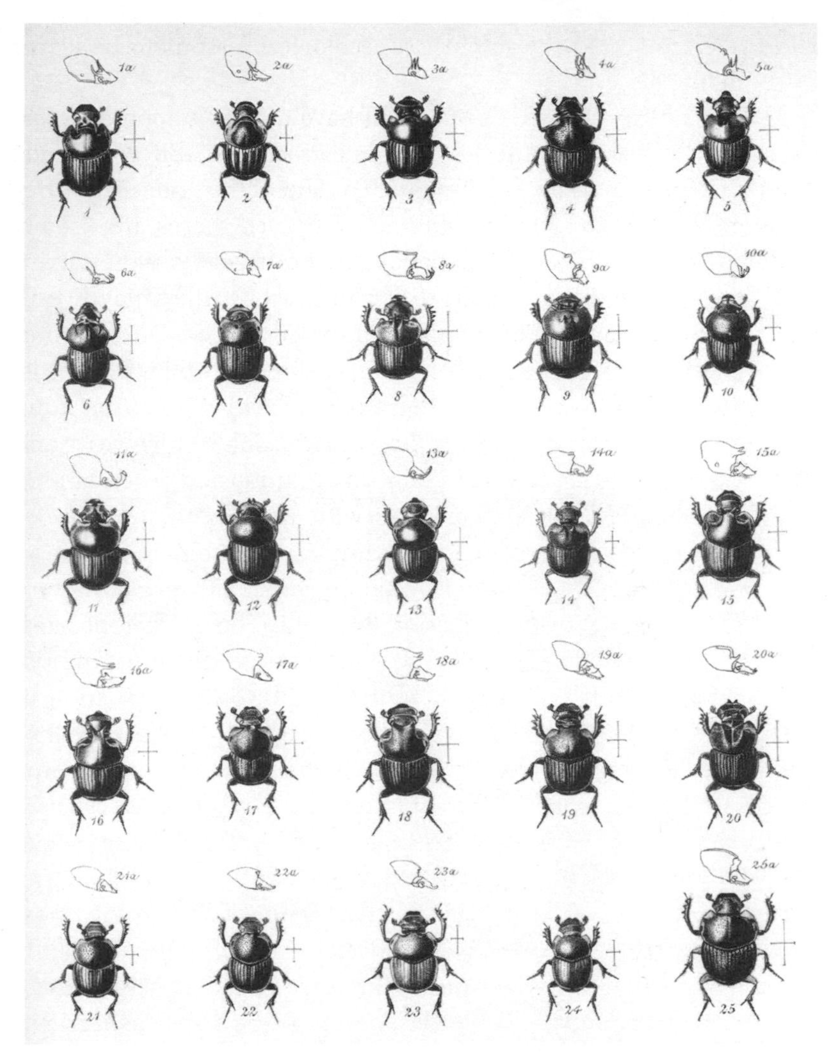

Fig. 16 Engraved plate of *Onthophagus* species from the famous *Biologia Centrali-Americana* (Bates 1886–1890).

the horns and ridges now make full contact with each other. They can lock against one another, just like the spike-and-bar projections of a handheld bottle-opener prising off the cap from a recalcitrant beverage. These contests of strength can last 75 minutes.

In the tunnel, push comes to shove now, and leverage applied by the head and thoracic horns becomes paramount, as well as their exact shape and length. In the squat, stout, muscular frame of a dung beetle, leverage is a function of body size, and body size is a direct consequence of nutritional intake during the beetle's larval stage. Within any population of a given dung beetle species, body size can vary by a factor of two or more, with smaller individuals barely half the size of their relatively gigantic confederates, even within the same pat, and in the same group of siblings. This size differential is reflected in the relative sizes of the horns too. Along this spectrum of body sizes of a given species, from small to massive, horns also vary from non-existent to monstrous. And it's mostly down to food intake as a grub – how much they got of that all too precious commodity in the dung pat.

MAJOR AND MINOR LEAGUES — MINE'S BIGGER THAN YOURS

The exact amount of nutrition received as a larva depends on dung quality as well as quantity. It also depends on temperature; feeding, digestion, metabolism and development proceed faster in warmer seasons or warmer climates. Thus, there is no fixed formula that states a larva must feed on so-and-so amount of food, for such-and-such a time, then change into an adult. But, even in the face of a low-quality diet, a limited amount of food, poor weather or northerly location, a larva can't just go on morosely eating and eating until it has finally had its fill. It may miss a vital synchronisation window with potential mates, it may emerge too late in the season to find suitable new dung, and the longer it goes on being a soft vulnerable larva, the more it exposes itself to the dangers of predators, parasites and disease. There is a pressure to get on and become adult, no matter how large or small you may be as a larva. At some point the decision to become an adult beetle, no matter your size, has to be taken.

Surprisingly there is not a direct proportional relationship between full adult body size (i.e. larval food intake) and overall horn length. Certainly the largest males of a particular species have the longest horns, but in the middle of the body size range in this species, the horns of apparently similar-sized beetles can vary from nearly maximal to almost non-existent.

This plasticity is a result of how the horns are produced in the metamorphosing pupa. Unlike, say, a leg, which is clearly defined in terms of size and shape and proportions by fairly strict genetic control, the horn is highly variable. It starts as a simple flap of cuticular skin as the beetle's adult shape takes form inside the chrysalis. Clusters of cells detach and proliferate, folding as they become trapped under the larval skin, but then expanding as the final moult gives way to the pupa. The embryonic horn sometimes looks like a series of concentric wrinkles, telescoped inside each other. But the final decision to go ahead with a prominent spike can be deferred until the last minute, or even reversed. As metamorphosis progresses, the horns can be remodelled, in both size and shape. In some species, programmed cell death, under hormonal control, can completely remove a horn apparent early on in the pre-pupa (especially one on the thorax), so that the adult beetle emerges surprisingly hornless. This also gives rise to the occasional horned sport, where no horn was known before.

The appearance, or not, of a prominent horn is not completely random. There appears to be a cut-off threshold at which point, after feeding has ceased, the beetle metamorphosing inside the pupa suddenly feels it can invest, after all, in a nice show-off spike. This threshold is under some genetic influence, and isolated genetic populations show slightly different patterns; thus in North Carolina most specimens of *Onthophagus taurus* (two bull-like head horns, obviously) over 5 mm wide have horns 4 mm long, whilst in Australian colonies they have to reach about 5.25 mm across the thorax before they can achieve such horn prowess (Moczec 2006).

Perhaps more surprisingly, a mid-sized dung beetle larva, changing into a mid-sized pupa, which might be expected to display a mid-sized horn, may actually emerge as a completely hornless male. This decision may be made right at the last minute, with potentially horn-enabled cells being reabsorbed and redestined, rather than being predestined from the outset. It seems there is some benefit from not showing off what would likely only be a modest tool after all, but in reallocating the horn material during metamorphosis to become a slightly bigger, if weaponless, rather effeminate, male.

These 'minor' males (as opposed to the brawny 'majors' with their swaggering body embellishments) nevertheless can get a piece of the dung action. They do not need to engage in loggerhead scuffles with other males. With their demure female-like hornless disguises, they can waltz right past, seemingly unobserved, and sneak-mate with the real females whilst the tusked warriors are otherwise engaged, battling it out with each other or shearing up the burrow walls and shaping the dung brood masses. Nutrition is all, but there are ways round any horn loss occasioned by not quite eating enough of the pie. The physically emasculated minor males can still gain sexual access, to pass on the genes that control, if necessary, horn reabsorption in the interests of false humility.

The clincher, if clinch were needed, occurs in two African species: *Heteronitis castelnaui* and *H. tridens*. These are the only known dung beetles where it is the females, rather than the males, which have the larger horns. In these species it is only the females that dig and burrow, and only they that remain in the tunnel to guard the brood balls of dung. Consequently, it is not the males that might need to fight over females, but females that fight over buried dung resources.

Fighting horns do appear in other groups. The aptly named minotaur beetle, *Typhaeus typhoeus*, is a dumbledor (family Geotrupidae), but is armed with three prominent thoracic horns in the males (small bumps in the female). Like *Onthophagus* appendages, they can vary from menacing spear-like prominences to rather feminine protruberances, and the assumption must be that they also fight, or sneak, in the deep tunnels they burrow. In Africa, where everything is bigger and more ferocious, males of *Aphodius renaudi*, are the only known examples of horn-bearing in this usually demure dweller genus of dung beetles.

Dung-rolling beetles will also fight with each other, but these are generally tussles over a dung ball, and here body size and leg length is everything. There is wrestling and pushing, flipping of heads and flailing of limbs, but without the constraining walls of the tunnel to allow locking of antlers, there is no tight head-to-head locking of weapons. Consequently roller males do not have horns. There's no point.

THE DOWNSIDE OF HAVING HORNS

It's not always easy having horns. The many bizarre horn shapes, their variation between individuals of the same species, and the useful identification features they provide to distinguish between species have led entomologists to invest much time in studying dung beetle appendages. For one thing, not all groups have them; as discussed earlier, they are most prominent in the genera that burrow in the soil under the dung, rather than those which merely wallow in faeces, or roll away balls of the stuff. They have evolved at least eight times

Fig. 17 *Onthophagus taurus* plate, from the seminal *British Entomology* (1823–1840) by John Curtis. The flower was merely an artistic juxtaposition – possibly more aesthetically pleasing for subscribers to the work than a cow pat.

in tunnelling dung beetles, but for each group with horns there are others, some closely related, lacking them. Having horns is not, in itself, a necessity for evolutionary success. Like many animal fight contests, real damage to each other is often avoided because one of the contenders very soon realises it will fail and gives up before blood (or haemolymph) is spilled. This usually takes only a few minutes. The other then celebrates victory by keeping or taking possession of the buried dung ball, the female, or both. But even though no body-wrenching clash of weaponry actually takes place, this doesn't mean that the horns become trivial – they are no mere costume pretence, they are very real and building them takes a real toll on the beetle's physiology. In fact, there is a significant biological cost to having horns (Harvey and Godfrey 2001).

Careful measurement of horn lengths within a species, and comparison with other body features shows that larger male head horns are associated with relatively smaller eyes and/or antennae (20–28% loss reported), as the limited resources within the developing pupa are traded off against each other (feeding has finished after all). Likewise, hornless females have relatively larger eyes than their horned consorts. Similarly, as thoracic horns increase in size, they are met with a comparative diminution in the beetles' wings (Moczec and Nijhout 2004). Here is a dangerous gamble (with life-changing consequences) made at the pupal stage, before the adult beetle has emerged. Anyone who has ever played computer games will recognise this choice when deciding on a character at the beginning of a contest – at its most basic this can be a choice between smaller and faster, or larger and more powerful. The dung beetle has to make this character decision without knowing what the world out there holds for it. The long-horned beetle major is hoping, perhaps, that pats will abound, that it will not have to fly too far on its slightly smaller wings, that its slightly reduced visual acumen and sense of smell will still allow it to find a suitable dropping. The rewards for the gamble succeeding are that when it arrives at the pat it will have a better chance of jousting its competitors out of the way, securing the brood ball and the female, and that the dung will be, as it were, its oyster. On the other hand, if the dung pats are sparser, a svelte hornless minor is, perhaps more able to complete its challenging quest on bigger, stronger wings, using its sharper eyesight and keener sense of smell to arrive at a distant dropping ahead of its clumsier, more obtuse fellows.

Perhaps the most remarkable trade-off is between opposing parts of the beetle's anatomy. It seems that increased horn size is directly correlated with decreased testes size. So those show-off bullies – you know, the ones with the biggest weapons – they really are compensating for something (Simmons and Emlen 2006). The evolutionary logic goes something like this. A well-endowed beetle can afford smaller testes because the beetle will easily achieve a suitable mate, and its sperm, though meagre, will be well targeted. After mating it will be capable of guarding the female so no further matings will dilute his sperm that she is storing. His small investment will be protected, and successful offspring will result. On the other hand, lesser-horned males may have to make do only with sneak matings, they must squander their seed, so their best evolutionary chance of success will be if they can use their copious sperm to mate with multiple females to try and sire at least a few offspring through sperm dilution.

BATTLING THE ELEMENTS TOO

It is not just with each other that dung beetles have to fight (literally and figuratively). They also have to battle the weather, most notably the heat. Insects, even large stout beetles, can only burrow into the dung, mould it, shape it, manoeuvre it and feed on it, when it is moist. As soon as it starts to dry out, the mushy fibrous mix starts to clag into a hard impenetrable mass with the texture of concrete (hence its suitability for a human building material), or else it becomes brittle and friable, disintegrating into dust and loose fibres. And as it dries out, so too the tell-tale volatiles that attract the dung inhabitants dry up; dry dung is no longer attractive because it no longer has the chemical scents that attract. This, of course, is another reason why there is the mad scramble to arrive at the new pat. Freshness is all.

On the whole, the coarseness of the dung's texture, and the precise fibrous make-up of the dung, a result of what the animal was eating before it voided, is of less consequence than the liquid and submicroscopic components in the moisture of the dropping.

Obviously it is the adult insects that arrive at a fresh pat, and despite the fact that dung beetles have chewing mouthparts, they are not equipped for much crushing, grinding and chewing of fibre particles; they concentrate on sipping the delicate juices of the dung.

From a neat series of experiments using various precisely engineered glass or latex beads of differing sizes, it is clear that even the largest adult dung beetles ingest only the smallest particles (8–50 μm diameter); they partake of the dung soup course only (Holter and Scholtz 2007). This is, nevertheless, highly nutritious, containing fermenting matter already half-digested, free-floating organic compounds and nutrient-filled bacteria beyond number. They lap this up as soon as they arrive, and despite my slightly facetious culinary quip about soup, this is a necessary and fortifying meal for any newly emerged dung beetle female. In the rush to get through larvahood, then manage the metamorphic change during pupation, and emerge as a fully functioning sexually reproducing insect, the resulting female's body size is just as influenced by the nutritional intake of the larva as the male's. But whereas a male beetle's horn size and shape is fixed and final, the female is able to extend the nourishment getting to her ovaries by continuing to take in food for herself. This maturation feed at the fresh dung is an important prerequisite before any thought of mating, nesting or egg-laying behaviour. After egg laying she will feed some more to mature a new batch of eggs ready for the next pat in a few days or weeks time. As the dung dries out, though, the soup becomes harder to find amidst the drying fibrous mass, and the dung beetles are unable to get enough nutrition. Eventually they give up and move on.

Flies do not have any jaws at all; their mouthparts are purely for sucking, so they too are only after the soup segment of the dung meal. It is not long before a pat, previously invisible under its bristling shroud of fussing and tumbling flies, is silent and bare. As soon as the outer layer dries and a crust forms the adult flies are unable to gain any more liquid nourishment. The occasional straggler may be able to find a crack in the crust, or explore just inside an entrance burrow of a beetle, but on the whole the dung is now left to the flies' maggots to consume from within.

Contrary to popular expectations, damp dung also attracts butterflies. Like flies, these are drink-only insects as adults, and although the coiled tubular proboscis is more familiarly thrust down into a flower nectary, many species have been recorded feeding on horse droppings, cow pats and dog dung. They tend to visit on hot days when the fresh dung is at its most fragrant, and perhaps when they are most in need of liquid refreshment. I've seen chalkhill blues, small tortoiseshells and a comma feeding on fresh dog dung, and although I know what they're after, it's always a novelty to see these pretty fluttering insects so seemingly out of place. In Britain the purple emperor is more often seen feeding on dung than any other foodstuff, mainly because it does not visit flowers at all. Instead of nectar it subsists, as an adult, solely on liquids obtained from animal droppings, carrion, fermenting tree sap, rotten fruit and muddy puddles.

In the tropics dense clouds of butterflies, frequently many different species together, settle on the wet mud at the sides of streams and ponds where cattle or antelope drink, and which have been liberally splattered with the animals' dung and urine. Such behaviour is called puddling, and the insects appear to be obtaining vital nutrients such as mineral salts, sodium and ammonium ions, amino acids and simple carbohydrates.

One of the most unusual instances of dung-feeding occurs in the Egyptian vulture, *Neophron percnopterus*, which occurs as various subspecies across southern Europe, Africa, Arabia to the Indian subcontinent. In Spain, the vulture is called *churretero* or *moniguero*, names which mean 'dung-eater', because of its conspicuous coprophagy. The normally meat-eating birds do not obtain protein (<5% in dung) or fat (<0.5%) from the cow, sheep and goat droppings, and the undigested cellulose would be alien to their digestive system anyway. They are after carotenoids. These are natural yellow pigments, mainly made by plants and bacteria, which the birds cannot biosynthesise themselves, and which they can only otherwise find in eggs or a few insects; these are unreliable sources where the vultures scavenge. Dung, however, is freely available, and contains the carotenoids the birds need (Negro *et al.* 2002). The pigments are important to the vultures, which have bright yellow beaks and faces, free of feathers, which indicate dominance status and are used in mating displays. Again, though slightly indirectly this time, you are what you eat.

MINORITY DUNG USES

Using dung as a food source is the major motivation for most animals' scatological interest, and this will be the central theme for the rest of the book. However, before concentrating attention solely on coprophagy, there are a few minority dung uses which bear scrutiny. These are not accidental dung uses, just those that have not quite made it into the mainstream yet.

Dung makes good bait. Wasps of various species regularly haunt fresh droppings, but they are not after the moist faeces themselves, more the other insects also being attracted – upon which they prey. The digger wasp, *Mellinus arvensis*, will sit atop a new horse dropping and pounce on the many dung-flies that are attracted. There's a video of one hunting around a badger latrine, available on the internet. Like all wasps, it's a predator, stocking its nest with the remains of its dead insect victims for its grubs to feed on. The nest burrow (well away from the dung) may terminate in a half-dozen cells, each stocked with 4–13 flies. As with many predators, once one attack is successful, the wasp learns from its triumph and returns to the same site repeatedly, to collect food, and it will revisit the same dung constantly throughout the day.

I once watched social wasps (yellowjackets) *Vespula germanica* trying to hunt down dung beetles arriving at a fresh patch of ripe dog dung on Hampstead Heath (Jones 1984). There were several wasps, and they made many unsuccessful attempts to grab the small mottled beetles, *Aphodius contaminatus*, which were busily flying in. Although they met their intended victims in mid-air, the wasps were obviously not quick enough to get in a fatal bite, before each flying beetle folded up its wings, closed its tough carapace wing-cases, and dropped into the grass to continue its mission on foot. And yet with six or eight wasps in attendance, they must have been having some success, presumably with the greenbottles and blow flies that were also being attracted.

The hornet robber fly, *Asilus crabroniformis*, rarely misses though. This large handsome brown and yellow insect, one of Europe's largest flies, is a fearsome adversary, and can bring down a flying dumbledor (*Geotrupes* species) with its massive spear-shaped mouthparts. Dung beetles make up a fair proportion of this robber fly's prey items, since it has a penchant for launching its attacks from drying cow pats. I don't think egg-laying has ever been observed in this species, but it is strongly associated with grazing meadows, where the predatory larvae live in the soil around cow dung, probably feeding on dung beetle grubs and dung-fly maggots.

Several types of beetle are also predators, but rather than using the dung as an attractant bait and waiting for prey to fly in towards it, they are more established as part of the in-dung community, so will be looked at in the next chapter.

The use of dung as a house-building material is, not surprisingly, also found in nature. The black lark, *Melanocorphya yeltoniensis*, and to some extent the sociable lapwing, *Vanellus gregarius*, have a strange habit of arranging a tumbled pavement of dried horse or cow dung pieces around their nests, on the grassy steppes of Kazahkstan and Russia (Fijen *et al.* 2015). This 'decoration' has produced quite some discussion amongst ornithologists. It seems to be something more than the birds merely using whatever available material they can find to create a suitable dish-shaped cavity on the ground to contain the eggs. They seek out the dung and use it in a defined way to create a closely laid patio around the nest. In the black lark, especially, the orientation and density of the dung pieces around the edge of the nest is predominantly coordinated in a north-east direction, suggesting that there may be some climatic or weather-related reason for the arrangements. One suggestion is that the dried dung pieces absorb the sun's heat during the day, and buffer the cold north-easterly winds during the night. A sort of night storage heater for the birds. Elsewhere the incorporation of dung into bird nests is common, and well known in blackbirds, thrushes and robins, who frequently line the inner bowl of their twig and grass constructions with the dry, soft vegetable fibres.

A particularly niche building application occurs in the larvae of various leaf beetles, which use their own droppings to disguise themselves. Most remarkable among these are the tortoise beetles. The adults are gently domed and flanged around the edges, very tortoise-like. They are able to hide their legs and antennae, retreating under the hard carapace of their thorax and wingcases if attacked, but their strange spiny larvae use an umbrella of their own stringy faeces under which to hide. The palmetto tortoise beetle, *Hemisphaerota cyanea*, collects a lifetime thatch of extruded excrement over its back, and looks more like a disembowelled ball of string than a living creature (Eisner and Eisner 2000). Other species are less flamboyant with their faeces, but even the common European green tortoise beetle, *Cassida viridis*, manages to collect a knob of brown and black frass which it holds over its back on the aptly named faecal fork – a long, pronged, hinged structure at its tail end. The advantages of your own dung parasol are probably twofold for the beetle larva: the dark twiggy strands make the creature look wholly uninsectlike, foiling any bug-shaped search-image used by birds, which are mainly visual hunters, to find prey; the dry brittle stalks are also unpalatable to these birds, which were no doubt hoping for a plump moist morsel, rather than an animated brillo pad.

Slightly less precisely engineered, but nevertheless equally effective, lily beetle larvae manage to completely cover themselves with their copious semi-liquid droppings, by virtue of having their anus halfway along their back. The shiny red-brown slug-like larvae become almost invisible in the rough jumble of slimy red-brown faeces that coat the leaves as these notorious garden pests shred the gardener's prize blooms. In the UK this is the black-legged *Lilioceris lilii*, but in much of Europe there is also the red-legged *L. merdigera*, from the Latin *merda* and *geros* meaning 'dung-carrying'. The excremental coating serves to hide the larvae from bird predators. Parasitoid wasps and insect predators are also deterred by the gluey mess, which gums up their legs, antennae and jaws should any of them be foolish enough to get too close. And I wouldn't be surprised if pesticide sprays are also deflected by the layer of dung insulation.

Several genera of leaf beetle in the subfamily Cryptocephalinae are called pot beetles, because the larvae live inside a hard or leathery pot made from their own frass, with just their front legs and head poking out at the front. As each egg is laid, the female beetle, dangling from a leaf, coats it with a layer of faeces (a process called scatoshelling), patting it down hard and even with her back legs before dropping it to the ground. The larva eats fallen leaves from the foodplant, adding to the rim of the pot as it grows, and only abandoning it when the new adult beetle emerges from its pupa. The pot cleverly hides the vulnerable grub, which just looks like a small particle of soil amongst the real soil particles. An even more advanced bit of coprocoating occurs in the closely related Clytrinae leaf beetles. Here the dropped frass-encrusted eggs are picked up by ants, confused by chemical signals from the larva, and taken back the nest, where the pot-laden grubs feed on dead leaf material and detritus in the colony.

Dung is a useful natural resource; it's readily available and versatile, but as is clear, there is one predominant use – as a foodstuff for all the scavengers that come along later. From now on, this book will concentrate on the key fact that animal dung still contains very high levels of usable nutrients, and that these are particularly attractive to insects – dung beetles and dung flies. One way of thinking about this is to follow their philosophy, and to regard dung simply as pre-owned food.

CHAPTER 5

DUNG COMMUNITIES – INTERACTIONS AND CONFLICTS

DESPITE THE SEEMINGLY mad free-for-all of new arrivals at a freshly deposited cow pat or horse dropping (or extruded stool), there is method in the insect madness, actually several methods. These are the various strategies that have evolved in beetles and flies (and a few other animals) to make sure they are successful in carving up just enough of the dung to get their offspring through to the next generation.

As was noted in the previous chapter, the first arrivers get a good start. And despite the early beetle appearances at my small private experiment around the back of the recycling centre near Reading, flies, by virtue of being good at flying, are often the very first to land. Indeed the horn fly, *Haematobia irritans*, is recorded laying its eggs in cow dung before defecation has finished. This is a rather risky business. Anyone who has ever witnessed cow dung coming into the world will know that its liquid texture means it is, quite literally, squirted down and splattered up. Being in the air near such a torrent is fraught with danger, and the small fly must calculate its trajectory and be aware of airborne droplets with expert precision if it is to dip down, lay its eggs and make its escape safely. It also uses horse dung, the drier boluses of which are, perhaps, less likely to splash, but which can still damage an overly keen insect unless it keeps a watchful eye on what is falling where.

The horn fly larvae are equally hasty; they hatch within hours and start feeding immediately. Within about 7 days they have devoured enough to pupate, ready to change into adults. They are still, however, in danger from predators, and must wriggle away from the pat to hide in the surrounding grass root thatch, but at least they have successfully taken their fill of the scarce dung resource.

A MODEL OF GOOD DUNG BEHAVIOUR

The yellow dung fly, *Scathophaga stercoraria*, is equally keen; in case we were in any doubt its scientific name means 'dung-eating dung-dweller', and it lives up to this by arriving on site seconds after the deposit has been made. These bright orange-yellow fluffy flies are not egg-laying, though, they are the males. They know the females will be along soon, but they are invariably first on the scene to carve out territories. Waiting for the slightly smaller, more demure, greenish-grey females to appear, there is a certain amount of aggressive posturing amongst the males. Like so many insects, larval foodstuffs

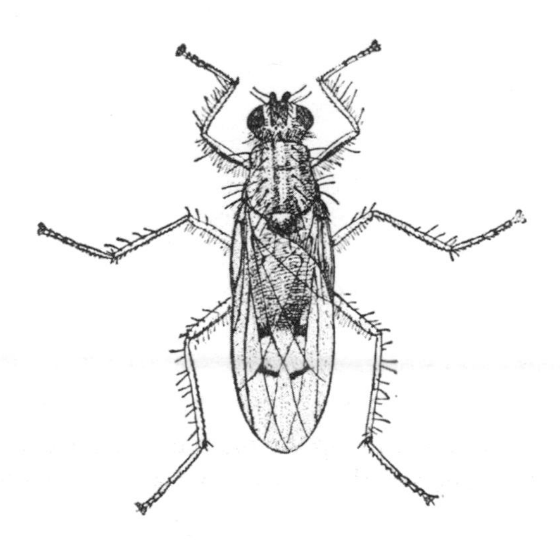

Fig. 18 Despite its bright pelt of hairs, the yellow dung fly avoids getting mired in the gloop.

and adult tastes differ widely; whilst the larvae may be dung feeders, the adult flies are ferocious predators, and they are not above using their powerful grasping legs and stout spiked mouthparts on each other in the mêlée. There is a tension in the air as a fresh pat 30 cm across may have 20–50 alert male dung flies each vying for its own precious few square centimetres of glistening faecal surface.

A ratio of four or five males to each female means that there is no such thing as a dung fly wallflower. The dearth of females may be because they are not able to mate straight after emerging as an adult; they spend 2–3 weeks maturing, catching and eating small flies and other insects to give them a protein boost in preparation for the arduous body-depleting business of laying eggs. The males, however, are ready to go much sooner. As each egg-ready female arrives at the pat, there is an undignified scramble. The nearest male grabs her and mates without anything in the way of courtship. A second or even a third male may pounce too, attempting to wrestle the female away from her first suitor, even if coitus is underway. Such is the competition for females that the successful male clings to his trophy wife when sperm transfer is complete to prevent the female mating again. Female dung flies can mate several times, but it is in the male's interests to hold on as long as possible, preventing latecomers intervening, thus diluting his sperm and reducing the number of his offspring amongst the eggs she lays. The female also gets some benefit from her protective partner, since constant pestering attempts by other males, desperate to mate with her, can cause physical injury to her delicate body. Still mounted on her back, he will stay in attendance as she lays her eggs in the dung. It is only as the last egg of the batch is laid, which the female signals with a sashay of her body, that the male finally relinquishes his hold and flies back onto the dung to await a new female arrival. Meanwhile the female flies off to feed on nectar and insect prey whilst she matures a new egg batch in her abdomen ready for another tussle on another cow pat in a few days' time.

The ease with which these flies can be observed on their pats, males and females easily distinguished, numbers counted, liaisons monitored and timed, egg batches measured, and paternity rates calculated, has lifted *Scathophaga stercoraria* to the status of model biological system. Scientific papers reporting on its behaviour, genetics, nutritional ecology and life history have burgeoned since G.A. Parker (1970) first took up the study. We now know quite a lot about the yellow dung fly; we can experimentally interfere with its life history and we can make quite precise calculations about its behaviour.

For example, as a fresh pat becomes crowded with males, the chances of a newly arriving male grabbing an incoming female falls, and it might pay a slightly late male to hang around in the nearby herbage, trying to intercept a female, rather than waiting in the main arena with all those other desperate competitors. There may be fewer females here, but this is offset by there also being fewer competing males. Very pleasingly the observed numbers almost precisely match the predicted numbers, with a large number of males on the dung itself, but the majority in the 20 cm zone of grass nearest to the dung, then numbers dropping off as they get further away (20–40, 40–60 and 60–80 cm zones).

Optimum mating and guarding times can also be calculated. It takes roughly 2.5 hours for any individual male to go through the whole process of finding a female, mating with her and standing guard whilst she lays her eggs. Given that the female may have mated previously, and still be storing some sperm in the storage duct (spermatheca) in her abdomen, the male needs to maximise his own sperm usage for her next batch of eggs. But the longer he spends inseminating her, the longer he puts off her actual egg-laying session, and the less time he will have to go off on his next romantic jaunt. The mechanics of fly sperm transfer can be measured by deliberately interrupting mating pairs at various times and seeing what proportion of the eggs are fertilised. It soon transpires that once a male can ensure 80% egg fertilisation with his genetic material, there is little to be gained from further effort – a neat example, from life, of the law of diminishing returns. This, then is the optimum time to stop mating and start guarding, so the female can offload her eggs, at least the majority of which he will be father to. The predicted time is 41 minutes of coitus and the observed time is 36 minutes; you don't often get such close agreement in biological experiments.

In my first year of university (1977), the week-long ecology field trip to Ashdown Forest had us studying birch invasion rates over heathland, the numbers of case-bearing amoebae living in the sphagnum moss, and measuring the arrival, mating, guarding and egg-laying times of cuddly dung flies as we lay belly down in a grazing meadow, closely observing fresh and fragrant cow droppings. Ah, such fond memories.

Scathophaga eggs are laid on the surface of the dung, in a crack or under the overhang of a wrinkle, but as soon as the larvae hatch they burrow down into the dung. For all dung-feeding fly larvae, speed is of the essence. Maggots are soft and vulnerable, easily attacked and eaten by other creatures (who'll be making their attack later on in this chapter), so they rely on being hidden in the dropping while they get on feeding as quickly as possible. *Scathophaga* takes about 3–4 weeks from freshly laid egg to newly emerged adult. Females emerge a day or so before the males, but need to spend another 3 weeks, feeding at flowers and eating small insect prey, before they are sexually mature. Sperm being easier to manufacture than eggs, the males only require about 1 week of predation before they are ready to mate.

Other flies are equally eager to get through the dangerous larval stage as quickly as possible. In many of the smaller dung flies (families Muscidae, Sphaeroceridae, Sepsidae, etc.) larvae have finished development in under 2 days, and a complete life cycle takes just over a week.

Even at this speed, though, competition is still strong; an early start does not guarantee enough food if everyone else starts early too. Just as with dung beetles, there is considerable size variation in dung flies, and if there is just not enough food to go round, a larva will take the gamble of becoming a small pupa, then a small adult, rather than waste time searching out leftover morsels in the surrounding soil. In laboratory studies of the common (and annoying) face fly, *Musca autumnalis*, larval densities were artificially manipulated and a threshold of 5 mg dry weight of dung was measured below which pupal and adult size fell away sharply.

MAKE WAY FOR THE DUNG MASTERS

Flies may be the flighty ones, getting in quick and racing through development as quickly as possible, but leave it to the dung beetles, the true masters of the faecal environment, to invent the most diverse series of fascinating behaviours to get around the constant problems of inter- and intra-specific competition. My personal beetle bias is going to come to the fore again here: although I will keep trying to introduce other groups, I maintain that a natural history of dung is really a natural history of dung beetles, and in particular those in the superfamily Scarabaeoidea. Others are with me on this, and there have been a series of hugely important ecological monographs in the last 25 years, notably those edited by Hanski and Cambefort (1991), Scholtz *et al.* (2009) and Simmons and Ridsdill-Smith (2011a); though highly technical and immensely detailed, they are a mine of information, and I have delved deep.

The Scarabaeoidea is a large group of attractive chunky beetles; it also includes the stag beetles (astonishing beasts) and the chafers (some of the most brightly coloured and beautiful insects on the planet). Dung beetles have long held a fascination for entomologists. Dutch naturalist Jan Swammerdam, writing his *Historia Insectorum Generalis* in 1669, enthuses of the two he had come across: 'One of them is conspicuous by a purplish gloss, like that of copper, on its breast and belly; the other glitters like green molten brass or copper delicately gilt, and indeed makes a very beautiful figure' (from the English translation, 1758). I hope my understandable fondness for them will be indulged. We know a lot about scarabaeid dung beetles; originally their relatively large size and attractive body forms made them popular to study (all those weird horns helped), they were easy to find (just look for a dung pat), and they carried out the seemingly driven, thoughtful, calculated behaviours of digging holes, excavating tunnels, burying dung and playing with balls of the stuff.

Once we get over the sheer amazement of just what these dung beetles are doing, the question soon arises: why? The answer to this question is now a vast field of entomological endeavour. The evolution of many of these behaviours appears to have been driven by the biological success of parental care, particularly in the complex procedures of making a nest. Most insects have no truck with parental care; large numbers of eggs are simply dumped somewhere and offspring are left to fend as best they can in the bug-eat-bug world out there. Thus, any parental care systems in insects are bound to attract the attention of biologists, not least because of their novelty. Parental care, and/or nesting, can really only work with manageable

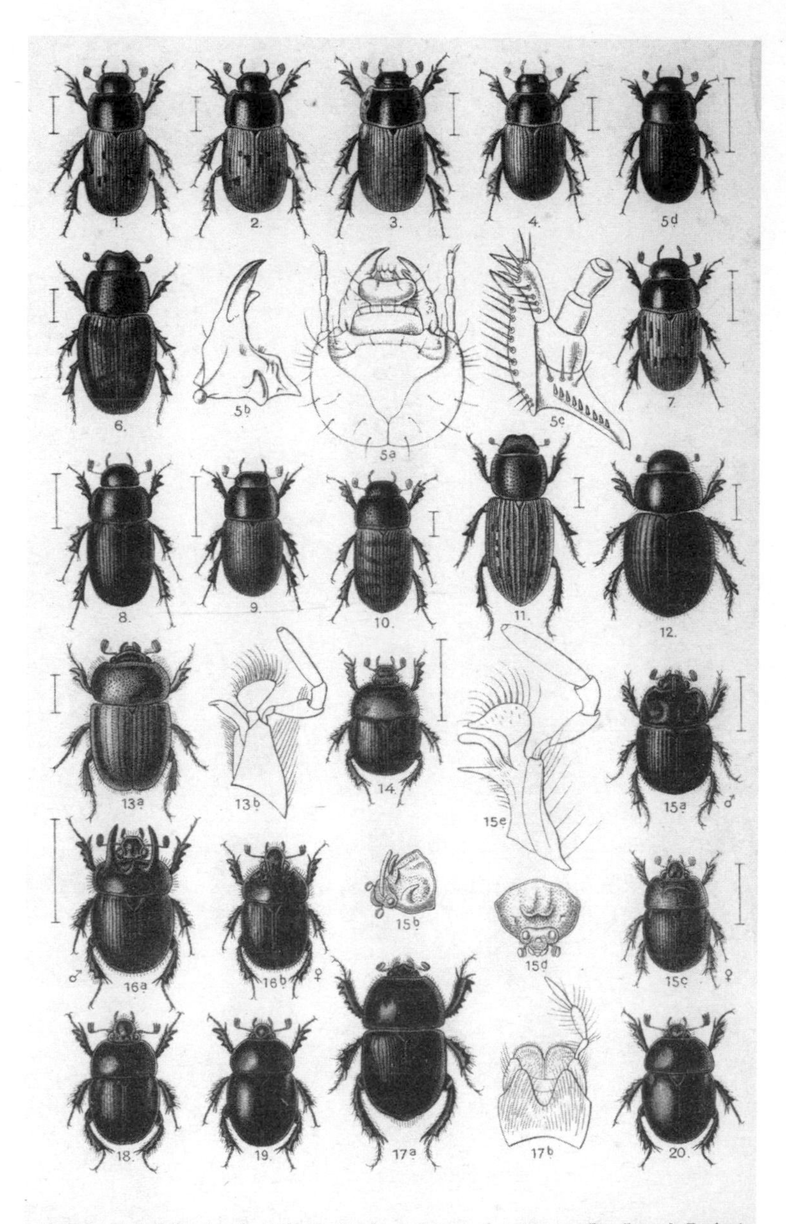

Fig. 19 Typical central European dung beetle diversity, from Reitter (1908–1916).

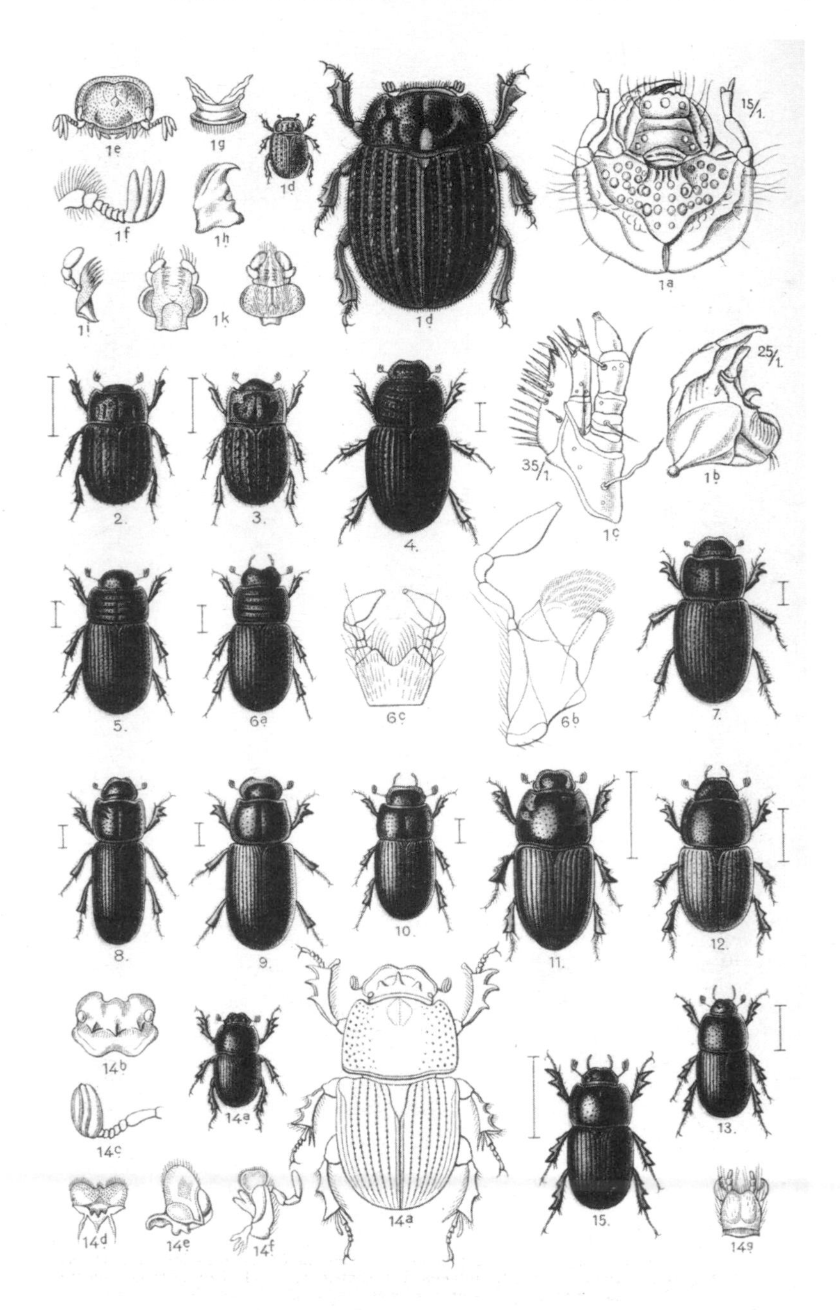

1. **Trox** sabulosus, 2. hispidus, 3. scaber. 4. **Psammobius** porcicollis, 5. sulcicollis. 6. **Rhyssemus** asper. 7. **Diastictus** vulneratus. 8. **Pleurophorus** caesus, 9. sabulosus. 10. **Oxyomus** silvestris. 11. **Aphodius** scrutator, 12. erraticus, 13. subterraneus, 14. fossor, 15. var. silvaticus.

(i.e. small) numbers of offspring. On the whole, dung beetles rear relatively few young, sometimes only one at a time. Nevertheless, the idea of dung beetles working to build a nest home in the putrid morass of the animal dropping might, at first, seem an unlikely suggestion; however, there are genuine analogies between nesting dung beetles and the better-known nesting behaviours of birds and mammals. A nest does not need to be an elaborate construction built of leaves, twigs, mud daubs, or other collected biological or inorganic debris. It can simply mean a hole (e.g. woodpeckers), a scrape (hares) or a tunnel (badger), or simply a small corner made one's own. The key attribute is that there is some sort of provisioning, either in terms of food, nesting material, insulation, security or protection from the weather. Now a nest in the dung sounds a bit more plausible.

It's all to do with reproductive pay-offs in terms of construction effort and the number of offspring successfully surviving. Without going into too much complexity here, it's enough to know that mathematical models can be designed to demonstrate how various behavioural strategies lead to various outcomes in terms of clutch size and number of offspring. There are several possibilities: (a) both parents can desert, the non-nesting strategy typical of most insects; (b) the male cares exclusively (rare, but does happen in a few birds and mammals); (c) the female cares exclusively (perhaps the most common, especially in 'higher' animals); (d) both parents care together. That last option is the one familiar to anybody watching a pair of birds working in concert to build a nest, lay and hatch eggs, then feed the constantly open craws of the hatchlings. Surprisingly this is also a common parental care strategy of many dung beetles.

Not all things are equal though. Traditionally, males and female animals are seen as having a different commitment to their offspring. The female, with a relatively huge egg investment, contrasts with the male's easy and copious sperm offering. An obvious male strategy is to mate with as many females as possible in the hope of siring as many offspring as possible. This is the tactic of those hornless minor males. This is all very well, but it immediately leads to sperm competition and dilution effects, and the possibility that, in the end, only a few males ever get to father anything. As with the dung flies, staying with a mate, and guarding her, can be the better option.

An important biological concept here (a central tenet of the trivial-sounding, but mathematically robust, 'game theory') is that parental behaviours evolve because of what is happening in the majority of the population, not necessarily in every individual. Infidelity and desertion may still occur, and strategies may change if circumstances change. There is room for behavioural plasticity. In the scientific jargon, an evolutionary stable strategy for biparental care can occur if the success of both parents working together is mostly twice that of one parent working alone. In simplistic terms, and slightly patronising tone, the family is greater than the sum of the individual parents.

In birds and mammals female-only care is very frequent, and in birds there are occasional examples of male-only care. In dung beetles a lone female can rear her own offspring, but since most of the 'care' is in the provision of a brood ball, before egg-laying, a male can only do so much on his own – he can't lay the necessary egg; male and female working together is the dung beetle brood-care norm.

Because dung beetles are easier to rear, easier to observe, and less ethically challenging to interfere with, they are now important biological models for testing nesting theories. Move over dung fly mate-guarding experiments, the beetles are here.

One easy measurement to come up with is the fact that the amount of dung set aside as a brood mass by the mother has a direct influence on the size and emergence of the offspring. Experimental manipulation of a brood mass, removing or adding material, is relatively easy. As is the measurement of any extra help in this, given by the male. Using offspring body size and male horn length as a good measure of larval nutrition, only the very largest females can rear handsome horn-toting major sons if they are working alone; but when a male helps, even the smallest mothers can produce majors. They are so proud. Interestingly, paternal help does not increase the number of brood masses (with one egg in each); this is limited by a finite number of eggs in the female's ovaries. Thus, male help does not increase the number of offspring, it increases their size and strength – their fitness in Darwinian terms of best being able to pass on their genes (half of which are his) to the next generation.

CARVING UP THE DUNG PIE – THREE FEEDING AND NESTING STRATEGIES

Broadly, there are three main nesting strategies for dung beetles (that's the large families Scarabaeidae, Aphodiidae and Geotrupidae again) using dung, and these all boil down to how the adult beetles deal with the dung when they arrive so they can best provision a nest, or at least secure a food store, for their grubs.

The most dramatic, the most obvious and the most often seen in wildlife films are the rollers – the telecoprids (from the Greek, more or less meaning long-distance dungers). These are the beetles which cut away a chunk of the pat, shape it into a ball and roll it away to bury it some way off. The ancient Egyptians revered the sacred scarab for this amazing behaviour, and it is similar species which are partly responsible for the wild frenzy when it comes to reducing a huge elephant dropping to meagre remains.

Next are the tunnellers – the paracoprids (more Greek, meaning beside, near or against the dung). They dig their burrows underneath the pat, or immediately adjacent to it, and drag balls, boluses, nuggets or pellets down into the soil, making small accumulations at the ends of the tunnel or in blind side-burrows. These beetles tend to work unobserved, but because of their head-to-head face-offs in the

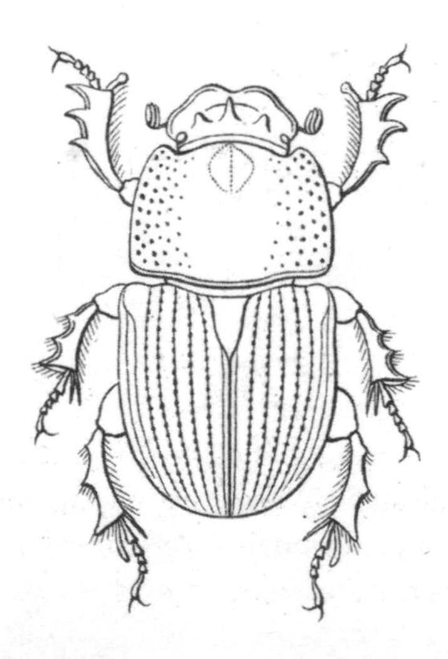

Fig. 20 Bold and stout, *Aphodius fossor*.

subterranean workings these are the species that have evolved the sometimes enormous, sometimes bizarre, head horns and thoracic horns.

Lastly there are the dwellers – the endocoprids (obviously meaning insider dungsters). Some of these just crawl or burrow into the dropping, chewing out or shaping voids, but not really doing much in the way of apportioning or concealing dung for the special use of their own offspring. Others gather a parcel of the dung and shape it into a discrete brood ball within the greater dung mass.

Taking these in reverse order, we'll start at the most fundamental dung–beetle interaction – the beetle just arrives at the dung and feeds on it, or lays its eggs in it. Arguably this is the most evolutionarily basic starting point for the complexities of nesting behaviour.

DWELLERS – AT HOME IN THE MIDDLE OF IT ALL

OK, I'm going to backtrack a bit here because in northern Europe the majority of true dung beetles are dwellers, but not necessarily nesters. The hugely diverse family Aphodiidae, mostly members of the genus *Aphodius*, are the majority stakeholders here. There are around 1,650 worldwide species in the genus, they tend to be smaller, narrower and more cylindrical than the other major groups, and they don't really make nests.

The aphodiids dominate dung in northern temperate zones. Even in Britain we have 55 species (compared to 15 tunnellers and no rollers), and of these, 15 species can be living together in the same cow pat or horse dropping – though the most I've ever found is 12. It seems they live a more leisurely life, and that here at least, they do not need to take part in the undignified scramble for possession. There is enough dung to go round, it does not dry out too fast in the cooler, damper climate, and anyway they are happy enough with slightly stale dung. They are mostly small – *Aphodius fossor*, at up to 12 mm, is the biggest, but most are below 7 mm – and they have adapted to the meagre and well-scattered droppings of deer, rabbits and the occasional entomologist obeying the call of nature. They tend to push underneath the dung, from the side, and will either burrow into the mass, hollow out small cavities or live underneath the pat. Here most lead an itinerant life, with little or no social interaction, and with scarcely anything that could be considered nesting behaviour.

Other dwellers, such as the scarabaeid *Oniticellus* (six Eurasian species), are less indolent. Although they remain in the centre of the dung pat, they actively shape portions into individual brood balls and lay their eggs in them. In true nesting spirit they may even remain on site to guard and monitor their developing offspring. Exactly where one brood ball begins and the dung mass ends is sometimes difficult to interpret, but some dwellers seem on the cusp of evolving more advanced nesting behaviour. The common *Aphodius erraticus* digs a shallow depression into the soil 3–5 cm deep, barely worth being called a tunnel, directly under the dung, and fills this with morsels of dung after laying an egg at the bottom. Elsewhere *Aphodius luridus* lays its eggs at the dung–soil interface and the larvae dig themselves short rudimentary tunnels into the soil which fill up with dung-infused liquid sludge every time it rains or there is a heavy dew.

As well as the 'true' dung beetles above, there are plenty of other dung-feeders from different beetle groups that also just feed inside the general dung mass. Some of these will be visited in chapter 6. Nesting is not shown by any of these other dung-dwellers, but it's important to remember that they and lots of other organisms are part of the scramble for possession, and it is this that has led to the evolution of slightly more complex brood-care behaviours.

TUNNELLERS – IN A HOLE IN THE GROUND THERE LIVED A BEETLE

The shovel-headed bulldozer-like form of the broad scarabaeid and geotrupid dung beetles couldn't be better designed for moving earth. They immediately set to work digging down into the soil, either right under the dung, or with an entrance close beside it. How far they dig down depends on the beetle's size and the soil type.

When, as a 14-year-old schoolboy, I helped out at an archaeological dig on the South Downs behind my parents' Newhaven home, we cut away the turf in rolls, and removed the meagre few centimetres of topsoil, looking for Iron Age post-holes in the underlying chalk just beneath the surface. We regularly found the duck-egg sized brood balls of dor beetles (*Geotrupes* species), 10–15 cm down; that was as far as they could go, before they met the impenetrable limestone bedrock. On the other hand I knew not to bother trying to unearth the minotaur beetle (*Typhaeus typhoeus*), which in the loose greensands of Ashdown Forest could easily penetrate more than a metre down.

It's not the kind of crass fact that dung beetle researchers like to brag about, but I've had a brief scour of the literature to find the deepest recorded dung beetle tunnel. The best I can offer is a North American species – the aptly named Florida deepdigger scarab, *Peltotrupes profundus*, with a burrow recorded by Henry Howden (1952) down at least 9 feet (2.7 m). There are just two caveats here. First, it seems that *P. profundus*, though obviously a dung beetle (family Geotrupidae), is not necessarily a dung-feeder, but a general soil humus and leaf litter detritivore; the non-dung foodstuff of some species is an important consideration when examining the evolution to and from coprophagy in chapter 7. Secondly, someone out there is bound to know of a deeper digger, so if they let me know there may be further updates in a second edition of this book.

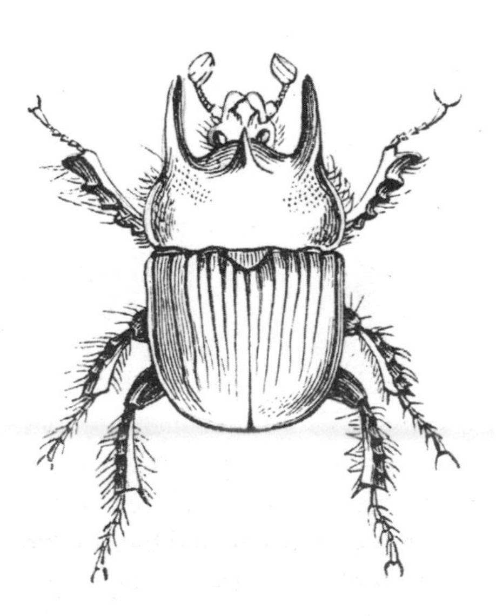

Fig. 21 Only the male of the minotaur beetle, *Typhaeus typhoeus*, has the three strong thoracic horns.

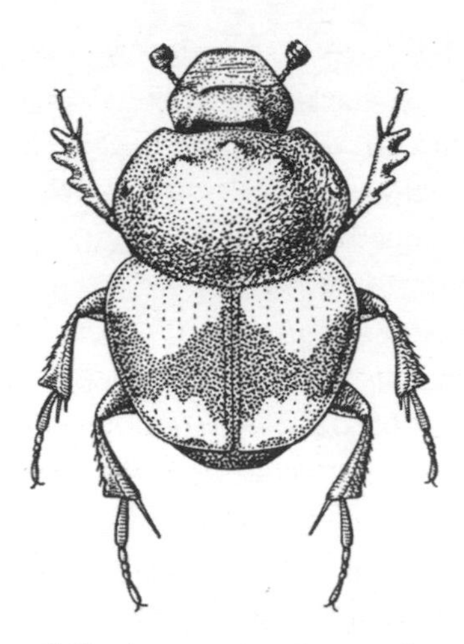

Fig. 23 *Onthophagus bifasciatus*, an Indian and Central Asian species.

Most tunneller burrows are of the order of 20–100 cm deep. The obvious point of burrowing is to get a portion of the dung out and away from potential competitors quickly so that it can be put to the sole use of one beetle; or a pair of beetles. Away from the mad scrambles on tropical savannahs, where hundreds or thousands of beetles jostle, but then each end up with precious little to their name, tunnellers can be the big-time winners. A pair of large tunnellers can easily bury 500 g of dung under a pat, so these beetles make a powerful contribution to dung removal from grazing pastures.

There is no hard and fast rule, but very often a male and female

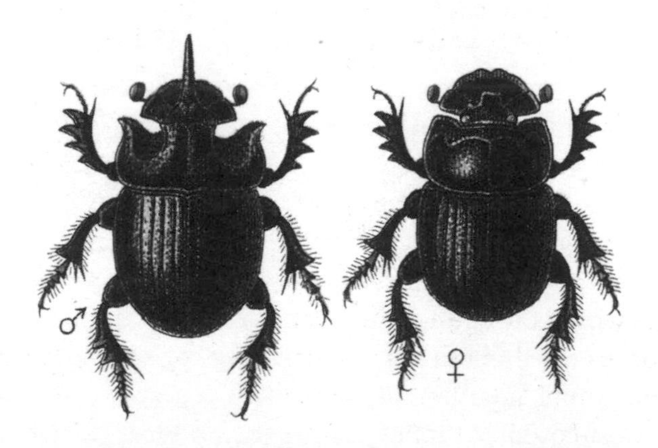

Fig. 22 Male and female *Copris lunaris* work together to raise a faimly.

meet at the dung patch, form a short-lasting bond and work together to build a nest. Each species will have its own strategy for where, how deep and how often it digs its nests. Here is where game theory predictions of paternal assistance and laboratory measurements of brood ball size come together.

As ever, in nature, there is often a trade-off between effort and results. The deepest and longest tunnels may be the safest, but they take time to build, and if the dung supply is limited it may be gone before an elaborate subterranean chamber can be constructed; or the beetles may only get one chance at it. A series of short tunnels can be dug fast, and stocked quickly with the diminishing dung, but they may not be far enough out of the way to escape interference and disturbance from all the other frantic diggers. Each dung beetle species adopts its own nesting strategy.

At its simplest, a female makes a rudimentary tunnel, sometimes just a vague depression, right underneath the pat; she buries a clump of dung, lays an egg and leaves it to get on with another one. This is the lowest-maintenance option, quick and easy for a female working alone, but still very basic. She will hope to have at least some offspring success through the larger numbers of eggs she releases, 10–50 per pat, maybe 130–150 in her lifetime. In other species a slightly larger hole is filled with several brood balls, all more or less touching each other, each with its own egg. There is less time digging, but this is still an easy option for beetles laying higher numbers of eggs.

Deeper tunnels are usually made in the species where males actively assist. These can be simple tubes, with a series of 5–15 brood balls, or sausages, wedged in at irregular intervals near the bottom. Like those produced by the minotaur beetle (*Typhaeus*), they may be quite deep, up to 1.5 m. The tunnel is dug first, then morsels of dung are brought down and packed around the egg. Subsequent eggs are laid in other balls as the beetles work back up from the deepest part of the shaft. Or the shafts may branch (an approach favoured by many *Onthophagus* and *Onitis* species), culminating in groups of short cells, into which the round dung balls are placed individually or in small clumps. At this level of sophistication one or both sexes may stay in the tunnel, guarding the brood.

The most complex tunnel-nesting behaviour is shown in a very few of the larger species. The 'English' scarab, *Copris lunaris*, and its

Fig. 24 Rollers hard at work, one of the Detmold paintings inspired by Fabre's works.

congeners, are good examples. Each tunnel is dug by the male-and-female pair and a small number (2–10) of large roundish balls of packed dung are created by bringing it down piecemeal and shaping it in the nest void. Each spheroid ball receives a single egg just inside one end. The nest chamber is large enough to accommodate all the brood balls, keeping them separate, and at least one parent (usually the female) remains in the nest, protecting and tending the developing grubs inside their spherical golfball-sized pods. She keeps the pods upright in the nest, and repairs them if they break open whilst the larva is developing inside. If a female is removed from the brood chamber, fungus soon develops on the dung balls. It is only when the larvae have achieved pupation, or are at least well into their larvahoods, that the adult beetle will move off to find another dung source and try to repeat the process. An adult female beetle may only live a year or two, and can only produce 10–30 eggs over her lifetime, so she invests whatever is necessary to give them the best chance of success.

ROLLERS – DIVINE INSPIRATION WAS JUST ABOUT RIGHT

Ball-rolling dung beetles show some of the most complex and fascinating behaviours anywhere in the insect world. It's no wonder that they inspired the ancient Egyptians to revere them, and to incorporate them into a complex other-world system of bizarre animal deities. At least one entomologist (Sajo 1910 quoted by Hogue 1983) has suggested that the biblical account of Ezekiel's wheels (Ezekiel 1:1–28) was a mystical allusion to roller scarabs, and not some new form of four-winged shining-bodied cherubim, nor alien astronauts.[1] Sadly, such rollers do not occur in Britain or northern Europe, but the original sacred scarab, *Scarabaeus sacer*, occurs in southern France, and was the inspiration for French entomologist J. Henri Fabre to write about the sacred beetle (1897). Coleopterists have been hooked ever since.

The ball is initiated by an 'active' partner; in *Scarabaeus* and *Canthon*, this is the male (but in other genera, e.g. *Gymnopleurus, Sisyphus, Phanaeus*, it may be the female). It may take a few minutes, to up to an hour, to scoop, shape and sculpt the ball, which is rolled only a short distance away at first. This initial ball may act as an attractant, bringing in a female to feed. Having travelled in some distance, possibly from her last nesting job, she will be hungry, in need of that ever so important dung soup to replace lost nutrients and replenish her ovaries. She is attracted by the dung, but is even more attracted by someone telling her they've got the dinner ready. In the closely related African genus *Kheper*, the male disports himself on the ball he has made; balancing head down, tail up at an angle of 45°, he releases a sex pheromone by striking his legs against abdominal glands, sending puffs of white powder, containing the scent, into the air (Tribe and Burger 2011). This is a clear indication to the female that a potential mate has made a head-start; she may feed on the dung ball, or get straight on with the business of putting it into a nest.

[1] I'm also partly swayed by his note that the Hebrew words for beetle (we still use scarab and carab today) and cherub (k'rubh) could have easily been confused over centuries of oral then handwritten copying and recopying.

When the 'passive' partner arrives, there may be some jostling, or she may be accepted immediately. There are few reports of anything that can be considered courtship, but it has been suggested that unequally size-matched beetles often fail to pair up. This may be because rolling a dung ball is easier for two equivalent beetles, rather than large and small struggling awkwardly together, like Laurel and Hardy moving that piano.

To roll a heavy dung ball, a dung beetle stands, head-down, on its front legs, and uses its long middle legs, and very long hind legs, to control the ball, as it pushes backwards, off and away. It is not just casually ambling along, easily trundling a smooth marble, but maintains a careful grip on the ball. If it tumbles it tries to keep hold, picks itself up and sets off again. In an odd departure from normal beetle anatomy, members of the genus *Scarabaeus* do not have front feet. The front legs, on which they push down onto the ground when rolling, lack tarsal segments and claws; however, the middle and hind legs have normal five-segmented tarsi and the usual double claw to maintain a grip on their sometimes wayward cargo.

As with tunnellers, there are various options as to how rollers manage their dung-removal strategy. Despite the obvious benefits of pair-bonding cooperation (after all, a dung ball can be 50 times as heavy as each individual beetle), there are rare instances of a female working alone. It is only regularly known in one species, *Megathoposoma candezei*, from Central America, the only instance of a roller operating without any male help ever. After mating near the dropping, the female sculpts out a ball of dung and rolls it away on her own, to bury it and lay her eggs.

The most common behaviour is for male and female to work together at the dung face. The female may spend more time shaping the ball, but the male is the powerhouse when it comes to rolling it away. They bury it a few centimetres below the soil surface, then mate. At this point they separate and the female works the brood ball into an egg or pear shape, maybe two. She lays an egg in the narrow upper end of each dung ball and leaves the nest to repeat the process, sometimes with the same male, if he is still about, sometimes with a new suitor. There is some variation to this behaviour. Some species of *Neosisyphus* have very long legs and may roll the dung well out of reach of other competitors, but they do not bury the brood ball;

instead they leave it attached to grass stems or twigs. This might save on time and energy, but risks desiccation and predation of the developing grub inside.

Further up the ladder of sophistication, *Kheper* females remain in the nest with the brood ball. The original rolled dung ball is reshaped into one or more (up to four, depending on the species) pear-shaped masses and each is inoculated with an egg. The mother then stays with her offspring until they emerge a month later. This is intense maternal care, and because of the seasonal movement of grazing animals, and the seasonal moisture content (or rock hardness of the soil), she may be limited to rearing only one offspring per season, and frequently only one per year.

Alternatively the bonded pair of beetles may return to the food source and bring further balls back to the nest. Eventually, when it is fully stocked, the female (sometimes the male too) remains in the nest looking after the brood, some species remaining in their nest until the new beetles emerge.

One intriguing aspect of brood ball nesting behaviour is that a grub feeding inside the bolus not only has a limited amount of food at its disposal, but is literally living inside its own food. This means it must defecate inside its own food too. As it feeds on the inside of the dung ball, it grazes out a spherical hole just a fraction larger than its fat, curled, C-shaped body. As it grows it must add its own frass back into the ball. By the time the larva is fully grown, it has eaten almost the entire contents of the ball, leaving just a thin shell 3–4 mm thick. This means that the larvae has ingested and reingested its own faeces many times over. This may be analogous to the caecotrophy shown by rabbits re-eating their night faeces.

The nesting behaviour itself is fascinating, but it is the rolling of dung balls that has captured the imagination of ancient philosophers and modern scientists alike. One of the most obvious things is that a roller (or a pair working together) do not just trundle off at random – they plot a trajectory and stick to it. In other words they set off from the dung in a straight line, and persevere no matter what gets in the way. If they meet an impenetrable and insuperable barrier, they may have to adjust direction for a short while, but as soon as they find a gap, or the end of the barricade, they resume their original heading. This doesn't necessarily mean that they are

Fig. 25 Condemned by Zeus to push a boulder forever uphill, Sisyphus, king of Ephyra, gave his name to a long-legged roller, beautifully depicted here by Detmold for Fabre's *Book of Insects* (1921).

heading for a known final destination; that would require some back-tracking triangulation when the way was clear again. Instead they are maintaining a fixed course, taking them away from the dung, away from competitors, as quickly and as directly as possible. There is no predetermined or favoured direction; mapping rollers moving away from the pat shows them radiating out to all points of the compass, but they are all moving in individual direct straight lines. They do this by using the sun.

There are some simple tricks to demonstrate this. Blocking the sun with a board, and using a mirror, moves the sun's apparent position in the sky, and the roller, striding out in mid-journey, quickly recalculates and reorients itself accordingly, unaware that it has been misled, and tricked into making an angle change. Conversely, gluing a small cap to the beetle's thorax, so that the eyes are covered, has them rolling their balls round in aimless circles and directionless spirals.

Dung beetles, like many insects (honeybees for instance) not only see light and shadow, and perhaps some colours, but they can tell the direction of the light as it passes through the air – they can measure its polarisation. Even if the sun is obscured by clouds, they can tell where it is, not by the relative brightness in the sky, but by the angle at which light strikes their eyes, something that we, as even very visually adapted humans, are incapable of doing.

The large African *Scarabaeus zambesianus* forages at dusk, and even though the sun is no longer in the sky it can detect light polarisation in the atmosphere, though human eyes struggle to make out anything in the gloom beyond a few dark blobs jostling in the dung. The uppermost dorsal ridge of its eyes contains huge light-sensitive cells, much longer and wider than in any of its day-rolling contemporaries. It is these which detect, not just much lower light intensities, but from which direction the light originated (i.e. where the sun is below the horizon). Looming over the roller's twilight perambulation with a large polarising filter, its axis set perpendicular to the recently set sun, immediately makes the beetle turn 90° left or right, realigning itself so that it thinks it is maintaining the same course. As it moves out from under the filter it rediscovers the sky's true polarisation pattern and reverts to its original trajectory. Entomologists are full of these experiments to tease the beetles.

Being active at night helps *S. zambesianus* avoid competition with much larger rollers such as *Kheper*, and on about 180 days of the year it can extend its foraging well into the night using the disc of the moon. Its visual acuity is so good that even if it cannot quite make out the pearly body of our celestial satellite, it can still detect the polarisation pattern, though this is less than one-millionth that of the sun.

Perhaps it should come as no surprise that dung beetles also use the stars to navigate. Marie Dacke and colleagues (2013) glued those same small caps over the heads of South African *Scarabaeus satyrus* and released them into a shrouded arena to roll their dung balls away as quickly as possible. They found that, on moonless nights, the light from the Milky Way was enough to guide the beetles. There is no suggestion that the beetles were navigating in the same way early pre-compass mariners used individual stars or constellations, but the diffuse band of pale light from the Milky Way did register, an observation confirmed in a similar arena set up in the Johannesburg Planetarium, thanks to assistance from the obviously patient and tolerant staff there. The latest suggestion is that the beetles, whether rolling by day or by night, take a visual snapshot of the sky, creating a mental picture of light intensities, polarisation, and position of the sun and/or moon, and use this stored mental image to compare to the physical world as they roll along (el Jundi *et al.* in press).

Whether travelling by day or by night, 20 minutes is an average ball-rolling time, allowing the beetle to get about 15 m away. The further it gets from the original pat, the safer it might feel about the food security of its offspring, or itself.

At this point I find myself slightly disappointed that nobody appears to have done a fastest-dung-roller Olympic speed trial. Again, perhaps that's not quite the done thing in a deeply serious scientific monograph. Anyway, I'd like to suggest the African species *Pachysoma gariepinum*, which charges around the Namib Desert at 0.33 m/s. That's a very respectable 1.2 km/h (about $\frac{3}{4}$ mph), and at least twice the speed of your average sacred roller. It probably does this to avoid burning its toes on the scorching sand of the desert's notoriously furious dunes.

Pachysoma is a bit of an oddity. It has no wings, and cannot fly. Indeed its wing cases are fused together down its back – an adaptation to avoid excessive water loss in this, one of the driest places on the planet. It gathers bits of dried dung and other detritus from the shifting dunes, making erratic wandering exploratory meanderings until it finds something to scavenge; then it makes a beeline back to its nest. It drags its booty behind it, instead of rolling it up front. Rather than setting off in a straight line from a large dung mass to bury its treasure far away from competitors, it ekes out a living in a bleak landscape, surviving only at low density, where really its only competition is with the fierce climate (Scholtz *et al.* 2004). What little it finds, it has to accumulate back at a base-camp, so it must somehow make a mental map of its immediate surroundings,

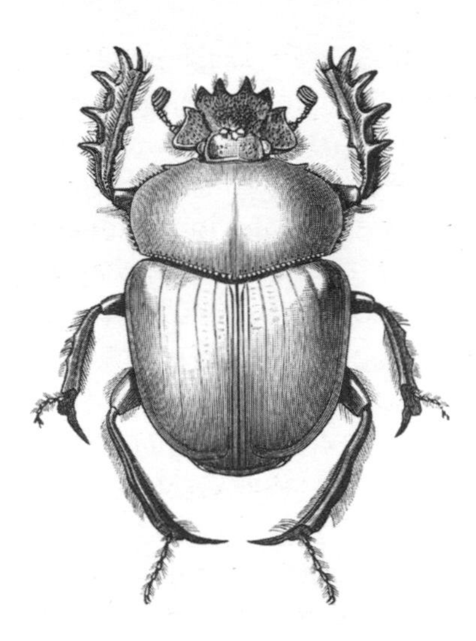

Fig. 26 *Scarabaeus sacer,* one of the largest, most powerful and most familiar of the large Old World scarabs.

presumably using rocks or clumps of dead scrub as landmarks to aid navigation back home.

THIEVERY – POSSESSION IS NINE-TENTHS OF THE NEST

Fighting is perfectly natural. It is, after all, a matter of life and death, at least for any potential offspring, whether a mate is wooed, a food is found, prey is caught, nest sites are secured or a big enough bolus of animal faeces can be assured. Fighting, or at least the threat of it, gave rise to those ridiculous horns in many tunnelling dung beetles. Subterranean contests go on, and there is usually a victor and a vanquished; very often it is sheer body size or horn length that determines this.

Rollers try and get their stash of dung as far away as possible, as quickly as possible: theft of dung balls is rife in the treacherous low-life neighbourhood of the dung pat. There are about 1,000 species of dung-roller beetle in the world, but in any given geographical location there is unlikely to be more than 10 (maximum 20) species. This is because competition is so intense that after massive squabbling there may not be enough for any single beetle to make one ball. Most of those thousands of dung beetles reducing an elephant dropping to nothing probably went off empty-footed. Speed may be of the essence, but so too is targeting a particular part of the day (or night), concentrating on a wetter or drier habitat, timing emergence to a different part of the season, or growing larger or smaller.

This last consideration is important, because size very often means might, and in nature might often ends up on top. On top of the dung sphere, a fighter will try to defend (usually) his ball against any other beetles trying to steal it. However, it is not worth a large roller stealing the insignificant ball of a small roller; it simply will not be big enough to rear a grub. This is good for diminutive species which might otherwise be the victim of much larger bullies. Fights usually occur between well-matched individuals, often of the same species. Both beetles grasp the dung ball and try to wrest it from the other's grip by making sudden flicking movements with their powerful front legs. In *Kheper* they might resort to head-butting and thrashing each other with their broad, spiked, front legs. Occasionally they will grapple together bodily, the loser eventually being tossed aside, literally thrown up to 10 cm away. The flattened front of the head, the clypeus, is used a bit like a frying pan tossing a pancake. Generally the larger beetle (whether the same or a different species) will win out and in a neat laboratory stadium experiment, nine diurnal and seven nocturnal species from Panama (Young 1978) showed a linear size hierarchy. In fight manipulations with *Kheper*, winners were on average 10% heavier than losers.

Occasionally contests involving three species are reported, when Jean-Pierre Lumaret observed *Scarabaeus typhon* make a dung ball in Corsica, it first had to fend off a similar-sized specimen of *S. sacer*; it was subsequently challenged by *S. laticollis* and the same individual of *S. sacer* returned for a three-way brawl. In the event, the original *S. typhon* remained the owner. A rightful owner retaining the ball seems the usual outcome in equally matched fights, the ball creator manoeuvring itself between the dung and the attacker, whilst continuing to roll the ball away.

Muscular power is not just linked to size, but is also related to body temperature. Dung beetles, like all insects, are poikilothermic (what used to be called cold-blooded): they cannot readily control their internal temperature (unlike you and me, homoeotherms). Mostly this means that a beetle's internal body temperature is only a degree or so above ambient, so insects have to wait until the day warms up before they can get going. Some bask in the sun, but they are very much at the mercy of local weather conditions. In the cool morning, when large elephant droppings are usually dropped, dung beetles need to overcome any dawn sluggishness, and get warm enough for active flight – around 34°C. Larger beetles can do this by shivering. This involves rapidly vibrating the internal flight muscles, without actually flapping their wings, generating physiological heat in the musculature. This raised body heat not only helps the beetles get to the dung pile in the first place, but also assists them should any conflict arise. As well as significantly larger *Kheper* beetles most often winning experimentally manipulated ball fights, so too hot beetles (average thorax temperature 38.7°C) tend to win over cooler ones (average 35.2°C). They're all pumped up and ready to rumble (Heinrich and Bartholomew 1979a). It doesn't always work out though. Staging test fights between large African rollers often resulted in the dung ball being torn apart in the frantic vigour of the contest. The evil scientists had to trick the beetles into fighting over artificial balls made of clay, impregnated with elephant dung juice. Rotters.

A warm beetle is also a fast beetle, so a large *Kheper laevistriatus* hot from its flight to the dung is quicker at sculpting the brood ball, and faster on the off as it makes a dash to bury it. The time required for a beetle 25 mm long to build a dung ball the size of a tennis ball can vary from just over 1 minute to nearly an hour. On level ground, roll rates of 14 m/min have been recorded by hot (>40°C) beetles; by the time they have cooled to 32°C they were dawdling at only 4.8 m/min. Back in the Namibian death dunes, though, speedy *Pachysoma gariepinum* can easily outpace anything else foolish enough to venture under the midday sun.

A CUCKOO IN THE NEST

Sadly, the beetle equivalent of bringing home the bacon and securing it in the larder does not necessarily stop widespread pilfering. Just as minor hornless males can sidle past their well-endowed major competitors to sneak-mate with the females, so too nest parasites can take advantage of another beetle's buried dung store and sneak in to lay their own eggs. This cuckoo parasitism is, like the bird that inspired the term, a genuine usurpation of the host's hard work and often results in the host grub's death in favour of the kleptoparasite (from the Greek κλεπτες, kleptes, 'a thief').

Amongst the dwellers, where no very clear nesting or brood ball creation is visible, it is difficult to distinguish between a true kleptoparasite and an interloper just burrowing over to pinch a bit of dung. In Britain, for example, species of the large genus *Aphodius* seem to inhabit just the pat, feeding on their bit as best they can. There are a few reports that the large and common *A. rufipes* can be a cuckoo parasite in the buried brood balls of the large dor beetle *Geotrupes spiniger*, whilst the small rare *A. porcus* has (once? Chapman, 1869) been found in the nests of *G. stercorarius*. Both, however, can also be found above ground simply 'in dung'. This suggests that kleptoparasitism is not a necessary part of these beetles' life cycles, but that if they find dung buried by some other beetle, they will take advantage.

Away from the temperate northlands, tunnellers and rollers abound; here cuckoo parasitism is more often reported, and we can make better assertions that some species are genuinely adapted to the parasitic lifestyle. In some places 10% of all dung beetle species are kleptoparasites. A single nest of the large tunneller *Heliocopris antenor* contained 130 specimens of ten other dung beetle species, although six of these (four species) were thought to be non-parasites which had just been caught up in the tunneller's zeal and buried accidentally in the large dung mass. Even the rollers can't get away from them: one large *Neosisyphus* ball contained 37 individuals of six cuckoo species, including the aptly named *Cleptocaccobius*. Infestations can be high and brood survival can be significantly impaired; in one study 12% of *Scarabaeus puncticollis* nests were attacked by *Aphodius* brood parasites, and offspring rearing rates were reduced by 68%.

There are fly kleptoparasites too. Several species in the genus *Ceroptera* in Africa and *Norrbomia* in North America (lesser dung flies, family Sphaeroceridae) have only ever been reared from brood balls buried by dung beetles. The diminutive adults ride on the relatively huge roller or tunneller beetles, waiting for a chance to lay their eggs before the dung is finally interred.

Whether nest invasions are deliberate cuckoo infestations or merely casual visitors (or accidentally buried in the heave-ho scramble), the nesters know they must protect their brood supplies. If the roller detects that his or her dung ball has been compromised in the making, and has been invaded by endocoprids, it will abandon the task. In the mad flurry, at the height of the elephant dung beetle season, endocoprids can render a large pat useless to rollers in just 15 minutes. It's a difficult trade-off, but despite the pressing urgency to get a dung ball to a safe distance, slower beetles may sacrifice getaway speed, but concentrate on creating a more solid, better consolidated, pest-free ball. They have to remain vigilant, though; females of the large European tunneller, *Copris lunaris,* will attack and kill *Aphodius* larvae if they find them in the nest.

PREDATORS – WHO EATS WHOM?

Yes, killing, it crops up everywhere in nature, and is the keystone of every food-web imaginable. So whilst the dung flies and dung beetles are gently clearing away the mess, or even fighting roughly over it, there are plenty of predators attracted too. The hasty assault on a large elephant dropping has as much to do with avoiding being eaten by hornbills, guinea fowl and mongooses, as it does with getting a fair share of the pat. Not that a quick exit necessarily guarantees the safety of a beetle, or its progeny. Days or weeks after the work is done, buried dung beetle pupae are still dug up and eaten by ratels (also called honey badgers) and aardvarks. As was alluded to earlier, wasps and robber flies sit and wait on the dung for incoming prey, on which they pounce. And certainly dung flies, getting to grips with each other in mate-guarding and egg-laying bouts, are not above eating either each other, or the myriad other small flies hopping about. But there are also plenty of predation opportunities inside the dung too.

The rove beetles (Staphylinidae) are perhaps the most diverse beetle family on the planet, so it's not surprising to find some of them live in dung. Though a few of the smaller slow-moving species

(*Oxytelus*, and *Anotylus* species) are dung-feeders (as both larvae and adults), the vast majority found in the pat are predators. Some are very small, like the taxonomist's nightmare genus *Atheta*, with who knows how many hundreds of species worldwide, none larger than 5 mm long. They, and their larvae, attack tiny invertebrates, especially young fly maggots. Then there are the larger, feistier species, including the large genera *Quedius*, and *Philonthus*, which can handle nearly full-grown maggots, and the handsomely mottled *Ontholestes* which dashes about at top speed around and under very fresh dung, snatching blow flies and greenbottle flies out of the air as they fly into the fresh dropping. These are quite spectacular to watch, and sure enough at my dung timing experiment near Reading one of the UK species, *Ontholestes murinus*, appeared in the first five minutes and had caught and made off with a hapless blowfly within moments.

Among the most characteristic of the predators are the oddly globose, shining and slow-moving clown beetles (family Histeridae). Quite how they got their common name is a mystery to me; according to some sources, *Hister* apparently means 'actor' in Latin (hence histrionics), and their flat legs, perfect for burrowing through the rancid mire, are reputedly likened to flat clown shoes and ill-fitting trousers. Seems highly unlikely. These chunky lumbering insects are never very large (12 mm maximum), but this belies their potency as ferocious predators of fly larvae. They are powerfully built, strongly armoured and have sharp protruding jaws. Their larvae, too, wade in on the attack. Dung is home to many species, but since their fly maggot prey will also breed in carrion, compost, rotten fungi and other decaying organic matter, histerids too are fairly catholic in their habitat requirements.

Not all fly maggots take on a passive victim status. There are also plenty of predatory grubs in the dung. Some, like those of the large bluebottle, *Polietes lardarius*, and the noon fly, *Mesembrina meridiana*, feed on dung material when they first hatch from the egg, but as they get larger they become more predatory, using the extra protein boost to finish their larvahoods, ready for pupation and the change to adults. Some larvae are predatory from the start, including a whole host of small nondescript grey and brown flies in the genera

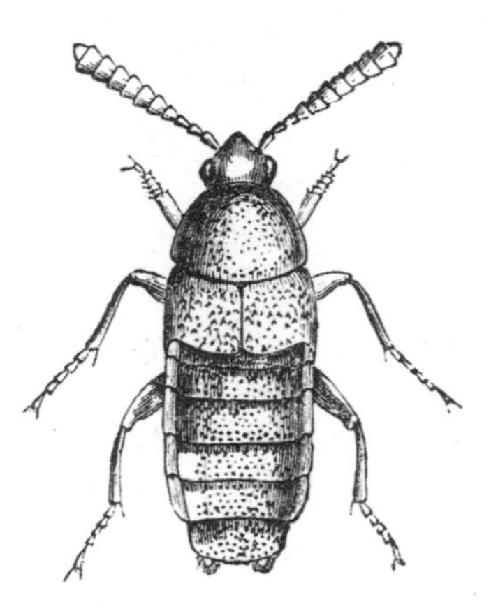

Fig. 27 The chunky form of *Aleochara* is useful for burrowing into dung, but the adults are rather transient, laying eggs then departing.

Helina and *Hydrotaea*. They may be diminutive, but there is always something smaller in there to get your teeth into.

PARASITES AND PARASITOIDS – THE ENEMIES WITHIN

A sadly understudied group of insects make up a poorly understood dark corner of the dung food-web – the parasitoids. These are mostly small or very small, slim wasp- or ant-like creatures in a broad range of families, including Braconidae, Ichneumonidae, Pteromalidae and Proctotrupidae; they are all in the same insect order (Hymenoptera) as bees, wasps and ants, but are only distantly related. Throughout all nature the parasitoids are highly numerous and massively important in the ecology of things, but such is their obscurity that there is no sufficient common name for them other than, collectively, parasitoid 'wasps'. Unlike predators, which attack and eat from the outside, parasitoids start internally.

The usual process is for an adult parasitoid wasp to find an insect larva and lay its egg on or in it. The egg hatches and starts to eat its host, alive, from the inside. The host larva may continue living and feeding for some days or weeks, but its fate is sealed. It may even get to pupate and form a chrysalis, but only the adult parasitoid wasps will ever emerge, usually through a neat round hole chewed in the side of the empty husk of its host. At some point the host is overwhelmed by the alien developing within, and it dies. Incidentally, that is the difference between a parasite, which lives upon but does not kill its host, and a parasitoid, which does.

One of the problems with studying parasitoids is that the adults are usually found well away from their potential hosts, sitting in the herbage, resting on leaves or even visiting flowers. Sit and watch a dung pat for as long as you like and I'm prepared to wager you'll never see one of them come near. It's all very frustrating.

One genus we do know a little about are the scarab wasps of the genus *Tiphia*. These are much more wasp-like, closely related to the myriad species which make small tunnel nests in sandy soil and stock their brood cells with flies, beetles, spiders, bees or other small insects, on which their grubs will feed. *Tiphia*, instead of digging the hole first, then stocking it with larval food, goes off in search of a beetle grub that is already buried, and digs down through the soil to find it. You'll still never see one on a cow pat though. It lays its egg on the grub or pupa and leaves its own maggot to get on with the gruesome demolition work, first sucking out haemolymph, then burrowing in and eating the viscera. Although *Tiphia femorata* and *T. minuta* parasitise *Aphodius* dung beetle larvae in Britain and Europe, they are also happy with non-dung-feeding relatives such as chafers (*Rhizotrogus* and *Anisoplia* species), which eat grass roots.

A large group of rather squat rove beetles in the genus *Aleochara* blur the boundary between parasitism and predation even further. The larvae are specialist attackers of fly pupae. They start gnawing on the outside, but then burrow in and complete their development inside the soon empty shell of the fly's puparium (swollen chrysalis). When the adult beetles emerge they leave a rough jagged hole in the puparium skin, contrasting with the usually smaller and neater holes of true parasitoids.

We know so very little about the majority of parasitoids of dung-inhabiting insects, other than a few random specimens, almost accidentally reared from dung samples. It is mostly fly larvae that are infected, but no doubt, like *Tiphia*, there are parasitoids of beetle grubs too. This is a glaring omission. Elsewhere in insect ecology the importance of parasitoids is central to our understanding of how ecosystems work. They are as much a pressure on insect larvae as are predators higher up the food chain. Dung beetles are special, but entomologists really need to get to grips with the parasitic Hymenoptera if we are to bring dung study into the mainstream. Hmmm, maybe that's just my wishful thinking.

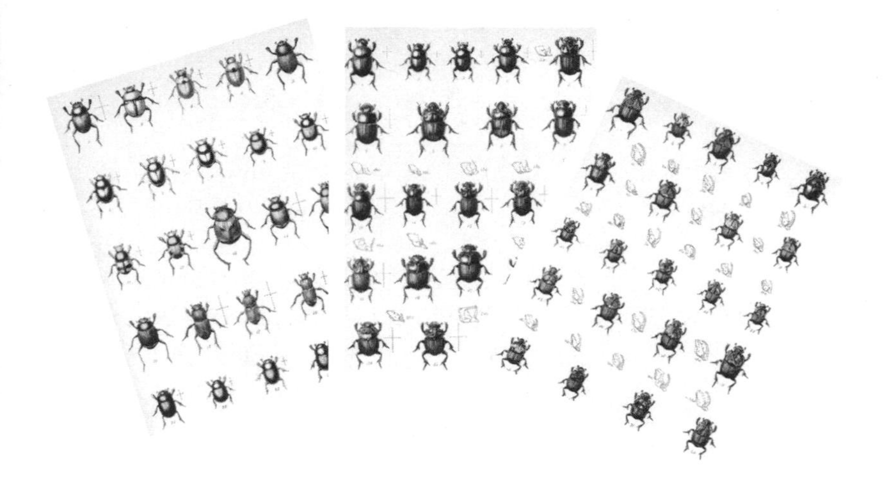

CHAPTER 6

THE EVOLUTION OF DUNG FEEDING – WHERE DID IT ALL BEGIN?

THERE IS NO doubt that coprophagy, the eating of dung, has evolved many times, in diverse groups of organisms. A quick run through the usual suspects shows that in temperate Eurasia and North America there are dung-eaters among at least five families of beetles and 28 families of flies (Skidmore 1991; Floate 2011). In the tropics entomologists are overwhelmed by species numbers, and observations become somewhat bleary at the edges, but as least 15 beetle families contain coprophages (my unpublished back-of-the-envelope calculation[1]) and, well, who knows how many dozens of fly families?

Everywhere there is dung, there are things eating it. It really is a widespread resource, not to be wasted. With such a rich and copious vein of nutritious material readily on hand, it is no real surprise that such a range of organisms have evolved to utilise it. Beetles and flies are the main dung devourers, but because they are so numerous, it is

[1] They are: Aphodiidae (dweller dung beetles), Scarabaeidae ('true' dung beetles), Geotrupidae (dor beetles), Hybosoridae (pill scarabs), Trogidae (hide beetles), Staphylinidae (rove beetles), Hydrophilidae (mud beetles), Elateridae (click beetles), Curculionidae (weevils), Ptinidae (spider beetles), Leiodidae (fungus beetles), Silphidae (carrion beetles), Agyrtidae (carrion beetles), Jacobsonidae (guano beetles), Tenebrionidae (darkling beetles). But I'm sure someone out there knows more.

easy to get lost in the detail. We can discover a lot about the evolution of coprophagy by looking at some of the odder outliers. Believe it or not, dung feeding also occurs in some cockroaches, crickets, termites, earwigs and moths.

That cockroaches eat dung is only to expected; they seem to eat anything. It's their propensity for hanging around the latrine, then nipping indoors to nibble on the breakfast cereal that has given them such a reputation as domestic pests, traipsing disease organisms all over our food. Yes, a few cockroach species have successfully invaded human habitations, but the vast majority of the world's 4,500 species are scavengers. They are a highly successful group, and have achieved this by maintaining a broad range of foodstuffs that they can devour, and by having toughened wing-cases (like beetles) for pushing through tropical leaf litter and humus soil layers. Here they grind away at every piece of organic matter they can find, including the occasional dropping. These are not really dung species, but the scavenging lifestyle has brought them into contact with dung, and this has allowed the evolution of a few specialist species that eke out a living in a bleak unfriendly environment, where there is precious little other than dung to eat – in caves. Cave cockroaches, *Trogoblatella*, *Spelaeoblatta* and others, live in dark subterranean caverns where they feed on the accumulated guano of bats, earning them the name guanobies. They have not fully let go of their precursor generalist tastes, though, and will also eat fungi, moulds, dead insects and any dead bats they happen to find too (Salgado *et al.* 2014).

The cave roaches are sometimes joined by cave crickets, *Ceuthophilus*, *Caconemobius* and others, descendants of the familiar singing and hopping meadow insects. Unlike their close relatives the grasshoppers (which are mostly herbivores), crickets are omnivores, predators and occasional scavengers. Deep in the caves, millions of years of evolution in the claustrophobic dark have made them pale, wingless and silent, and although some species retain long legs, the muscles have atrophied and they have lost the ability to jump. Like the guanobies, they eat whatever they can find, but bat guano is a mainstay of their diet (Lavoie *et al.* 2007). Bat guano is also an important nutritional source for the Alabama cave fish, *Speoplatyrhinus poulsoni*, and the grotto salamander, *Eurycea spelaea* (Fenolio *et al.* 2006), although these vertebrates probably feed on guano-feeding insects and worms to get the majority of their food. So far, no fish or amphibians have made a real concerted effort to evolve coprophagy.

In dry season Africa, termites are the chief removers of elephant droppings, which dry out quickly (Coe 1977). As demonstrated by the eager attentions of the sitatunga marshbucks, elephant dung is virtually untouched cellulose, and the termites treat it as if it were just chopped up dead grass or leaves, which of course it is. The fact that it has passed through the digestive tract of a large mammal is of scant concern to them. Although no jump to obligatory dung feeding has been made by any termite species, their involvement is not minor, nor random, nor accidental, and they are arguably making the same use of the dung as other dung-feeders; they are just waiting a while for the moisture content to go down a bit, and the faecal volatiles to lift.

You occasionally find a common earwig under a dried pat, but it is little more than sheltering there; however, the lesser earwig, *Labia minor*, is a true denizen of the manure heap. It used to be much more widespread in the UK than currently, and appears to have declined dramatically since the decline of horse-drawn vehicles. When one flew into my Dad's moth light-trap during the early 1990s, he reminisced that he had not seen one since his boyhood days in Shepherd's Bush, some half a century earlier. So much horse dung was dropped onto urban roads (hence the term 'road apples') until at least the first half of the 20th century that cleared heaps of the stuff dotted suburbs and city centres everywhere. The lesser earwig was right at home. Horse dung is relatively dry and fibrous, and parked on a flag-stone pavement or tarmac road it soon dries out and is not very attractive to dung beetles. It does, however, soon go mouldy, and it is the fungal mycelia of the moulds on which *Labia* feeds. This distinction between organic matter half-digested in mammalian guts, or half-digested by fungi and moulds, is a very fine one, and it is sometimes not possible to tell whether a dung-feeder is actually eating the dung material, the fungus decomposing it, or both.

The sloth moths (*Cryptoses choloepi* and others) offer a final window onto the evolution of dung-feeding. These small, slim, drab moths live in the untidy greenish pelage of several types of sloth, living in the tropical rainforests of Central and South America

(Bradley 1982). As in many moths, the adults do not feed, beyond a bit of dew-lapping, or whatever the equivalent water droplets are that form in their mobile carpet homes. About once a week the sloths climb slowly down from the canopy, scrape a small hole in the leaf litter with their short stubby tails, and defecate.[2] When they are done they smooth over the leaf litter and ascend again to continue their leisurely leaf chomping. However, in the few minutes they are occupied with the call of nature, some of the moths quickly disembark from their hosts and lay eggs in the dung, before hopping back on board. Their caterpillars will feast on the sloth excrement (Pauli *et al.* 2014). By default, moth caterpillars normally eat plant material, such as leaves, stems, flowers and seeds, but *Cryptoses* is a member of the Tortricidae, a moth family renowned for evolving a wide range of larval tastes; some of its members will feed on dead leaves, rotten leaves and fungi, whilst others have adopted a taste for feathers, fur, soft furnishings, carpets and silk underwear. Again, the switch from living leaves to dead, fungoid or mouldy leaves, to dung is an almost natural transition. Incidentally, the sloths derive some benefit from their hitchhikers. Inevitably the moths bring some biomaterial into the sloth fur, be this a few droppings of their own, or their corpses when they die. Minimal though this may be, it increases the density of the green *Trichophilus* algae which grows on the sloth's hairs, giving the slow hosts their camouflaged verdant colour.

The key evolutionary event in all of these odd dung-feeders was a switch from being generalist scavengers, possibly via fungal or mouldy decay, to finding nutrition in the form of partly digested plant material in dung. They then did well enough on it to become specialist excrementalists. In these cockroaches, crickets, earwigs and moths the switch came relatively late in evolutionary terms, with just the occasional scavenger species in these groups now eating dung. They have not yet had a chance to radiate much and therefore have not evolved a host of diverse new species. In beetles and flies, however, the switch to dung came very early, and this has allowed speciation at an explosive rate.

[2] Sadly, I can find no suitable technical term for sloth faeces, but can suggest either largos or adagios – these are, after all, slow movements.

THE GREAT BOWEL SHIFT

Flies (insect order Diptera) appear to have evolved from a common ancestor with fleas (Siphonaoptera) and scorpion flies (Mecoptera) about 280 million years ago (MYA), and many of the family lines still in existence today can be traced to about 200–150 MYA (Grimaldi and Engel 2005). Flies are easily the most ecologically diverse group of insects on Earth, but the majority of families have larvae (maggots) that are feeders, or scavengers in that catch-all substrate – decaying organic matter. Their success is often based on their ability to use whatever putrid matter is at hand. Breeding in dung is a natural progression from feeding on any other rancid material. Indeed, the notion that flies breed in dung is so deeply ingrained in the public psyche that most people assume that all flies breed in it. They can become quite defensive if you suggest any alternatives. Dung is highly attractive to many different groups of flies, but there are relatively few which have become so specialised that, today, they occur solely in dung.

The dung flies (Scathophagidae) are an obvious suggestion, but even here only a small proportion (5 out of 55 UK species, for example) are dung-breeders; the others are generalist scavengers, leaf- or stem-miners in plants, aquatic or terrestrial predators. Likewise the equally inappropriately named lesser dung flies (Sphaeroceridae) and dung midges (Scatopsidae) are mostly microbial grazers on decaying plants or fungi, with only a few of the better-known species being closely associated with animal dung. Among other fly groups, there are several hover flies (Syrphidae, the majority are detritus feeders or predators), a handful of soldier flies[3] (Stratiomyidae, most are aquatic scavengers) and a scattering of midges from various families – most, though, are fairly broad-spectrum soil-feeders. The tiny shining ant-like flies in the family Sepsidae are the most nearly

[3] The black soldier fly, *Hermetia illucens*, originally a North American species, is now deliberately cultured in manure and sewage management systems across the world, particularly on poultry and pig farms, where it speeds up decomposition and consumes competing larvae of house flies. Apparently the larvae are edible by humans, though this might require an almighty public relations exercise to bring this foodstuff into the mainstream.

entirely coprophagous group; the majority do breed in animal dung, but a few also breed in rotting seaweed.

Perhaps the only fly families solely adapted to faecal feeding are the Mormotomyiidae and Mystacinobiidae, but these contain just one very peculiar hyper-evolved specialist species each. The larvae of *Mormotomyia hirsuta* and *Mystacinobia zelandica* feed on bat guano and are known, respectively, from a single cave in Kenya (Kirk-Spriggs *et al.* 2011), and a few hollow trees in New Zealand (Holloway 1976). These are interesting, but ecologically not terribly significant.

One of the most important fly groups is the Muscidae, a very large family that contains many important dung-breeding species including: house fly (*Musca domestica*), bush fly (*M. vetustissima*), face fly (*M. autumnalis*), noon fly (*Mesembrina meridiana*), horn fly (*Haematobia irritans*), stable fly (*Stomoxys calcitrans*) and mottled flies (*Polietes* species). Even here, though, for every muscid species found breeding in dung, there are probably 50 found breeding in other rancid decay, in leaf litter, in soil, or they are predators. In a nice historical coincidence, the earliest description of a fly breeding in faeces was given in 1669 for the muscid *Fannia scalaris* by Dutch biologist and microscopist Jan Swammerdam in his *Historia Insectorum Generalis* (*General History of Insects*), who found it close to hand in human dung.

Again, only a few species amongst a whole raft of generalist scavengers develop exclusively in excrement, suggesting that the switch to dung has occurred somewhere recently down each evolutionary line. Nevertheless, because dung-breeding flies occur in so many different families, it is clear that this transition from general scavenging in decaying organic matter to faeces has occurred easily, almost inevitably, and many times over.

Things are slightly different amongst the beetles. True, there are a few groups where dung-feeding is in the minority. The 'swimming' dung beetles (*Sphaeridium* species), and their less streamlined, more domed relatives (*Cercyon*, *Megasternum*, *Cryptopleurum*, etc.) are part of a much wider group (family Hydrophilidae) which scavenge or hunt in muddy pond edges, and which have swapped swimming in water for swimming in liquid cow dung. And amongst the huge hyper-diverse rove beetles (family Staphylinidae) there are plenty which scavenge in rotting detritus, so no surprise to find a few dung-eaters in there too.

One of the strangest dung-feeding beetle groups occurs in Australia. Weevils, one of the most successful and diverse groups of beetles on Earth, are wholly plant-eaters, and have a longer or shorter snout at the front, with jaws at the end, to chew deep into plant tissue. Females generally have longer snouts to dig deeper to make egg-laying holes. Throughout the world many weevils eat dead plant material, including rotten wood, but only the Antipodean genus *Tentegia* feeds on dung (Wassell 1966). The two, or perhaps four, species (depending on which monograph you peruse) make small caches of kangaroo or possum pellets which they drag rather clumsily back to a small depression hollowed out in the soil under a log. The female chews the requisite drill hole and an egg is laid in each small pellet; the grub develops inside. This really is an example of dung simply being processed and repackaged plant material, and being reused accordingly.

These minor beetle groups are all very interesting, but, yet again, move over for the real dung beetles; here are three discrete groups of beetles (though all fairly closely related) which have so taken to excrementalism that they now dominate dung faunas throughout the world. The Geotrupidae (dor beetles or dumbledors), Aphodiidae (dwelling dung beetles) and Scarabaeidae ('true' dung beetles) account for the vast majority of beetles breeding in dung. There seems good reason to imagine that dung-feeding appeared in some common ancestor to these three major beetle lineages, or very early in their evolutionary divergence.

According to the fossil record, the Aphodiidae–Scarabaeidae split occurred about 140 MYA, and the implication is that the Geotrupidae branched off at some similar time. At first glance this might suggest that dung-feeding was already evolved at this point. However, things are never this simple.

It is clear to entomologists that these three groups, along with the stag beetles (Lucanidae), bess beetles (Passalidae), hide beetles (Trogidae), pill scarabs (Hybosoridae), bumblebee scarabs (Glaphyridae) and sand scarabs (Ochodaeidae) form a natural cohesive assemblage – the superfamily Scarabaeoidea, sometimes designated as the suborder Lamellicornia in older books. The entire

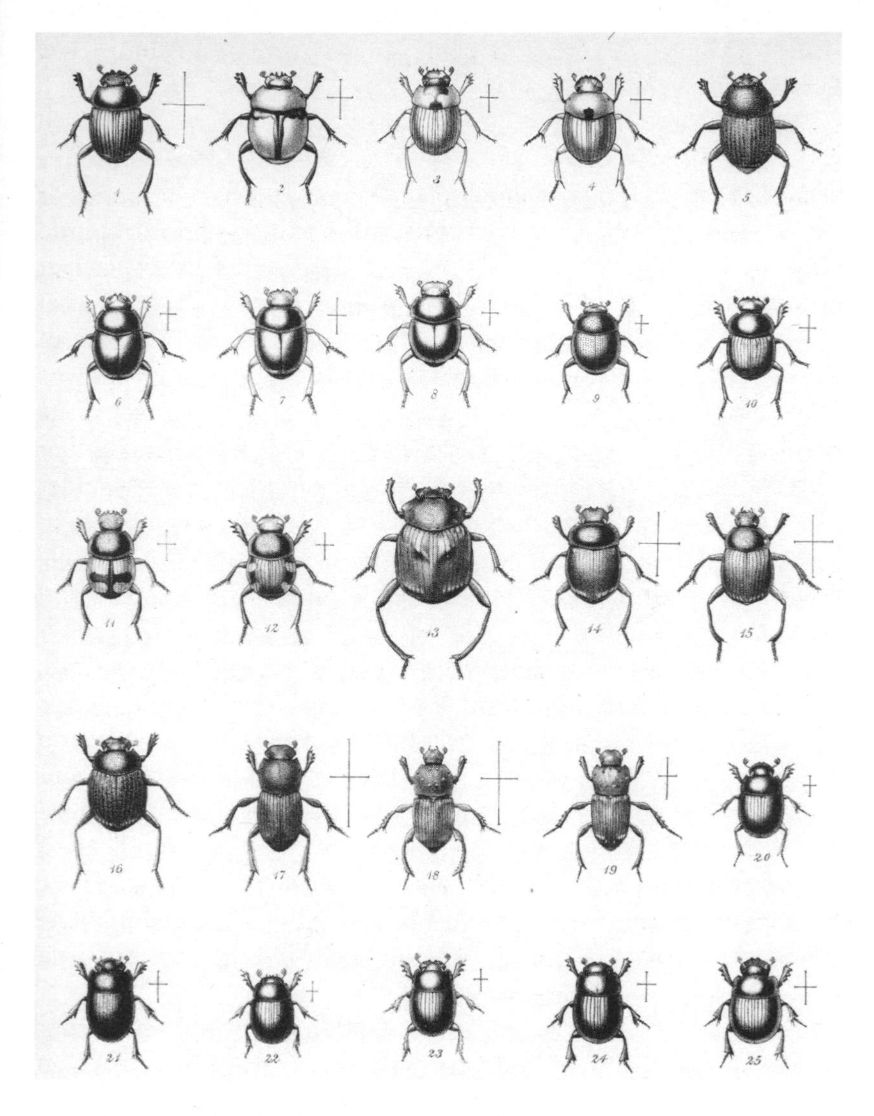

Fig. 28 Dung beetle diversity from *Biologia Centrali-Americana* (Bates 1886–1890).

superfamily Scarabaeoidea group have similar pale, squat, C-shaped grubs, and most of the non-dung-feeders today are soil layer and leaf litter denizens, eating dead leaves, fallen fruit, rotten wood, decaying fungi, roots or, you guessed it, other 'decaying organic matter'. They all, no doubt whatsoever, shared a common ancestor, but how they diverged into separate dynasties is still open to endless debate; check out Grebennikof and Scholtz (2004), Smith *et al.* (2006) and Philips (2011) for some broad friendly discussion. What is most confusing is that the family group containing the 'true' dung beetles, the subfamily Scarabaeinae, also contains the multitudinous chafers (subfamilies Rutelinae, Cetoniinae, Anomalinae), cockchafers (Melolonthinae) and rhinoceros beetles (Dynastinae), most of which are decidedly dung-averse. Larval forms, internal structures and DNA analysis all confirm that the 'true' dung beetles are more closely related to these non-dung species than they are to either the dwellers or the dors. So how and where did dung-feeding originally evolve in this group?

The jury is still out, and it seems that with each new study the family trees are redrawn slightly. When I set out to write this book, in early 2015, it was clear to me that the small cylindrical *Aphodius* dung beetles, which I had collected in Britain for 45 years, were all part of the same dung beetle family as the *Geotrupes* dumbledors and the squat *Onthophagus* tunnellers. But, oh no, other coleopterists with an unhealthy interest in cladistics (computer-generated classification schemes) had been busy rearranging all the beetles into new groups. Armed with overconvoluted computer programs, spanking new DNA analysis machines and too much time on their hands, they had thrown all my preconceptions out of the window. Then they started squabbling amongst themselves. I've just spent the better part of a whole morning deciding whether to treat the dwelling dung beetles as the family Aphodiidae (as suggested by some authors), or the subfamily Aphodiinae, part of the Scarabaeidae (as suggested by others). And are the dor beetles Geotrupidae, derived from the basal stag and bess beetle lineages? Or are they yet another subfamily of the Scarabaeidae? I don't know, and quite, frankly, I'm still confused. Whatever pedigree I use I can guarantee that some bright spark will come along tomorrow and rearrange things still further. It's enough to drive you to becoming a dipterist.

What is clear, though, is that a foodstuff switch (actually switches, more likely) from detritus-scavenging to dung-feeding must have occurred very early on in this ancestry, whichever half-baked tree-of-life scheme you decide to follow. It also seems clear that the first true scarabaeid dung beetle lived in South Africa. All the major dung beetle lineages in the group can be traced back to this region of what was then the southern hemisphere supercontinent Gondwana. Today the greatest diversity of true dung beetles (and also mammalian herbivores) occurs in the gently dispersing fragments of this antediluvian land – sub-Saharan Africa, tropical South America, Madagascar and Australia. Further north, the remnants of the northern supercontinent, Laurasia, have drifted to become North America, Europe, Asia and India, and although there are still plenty of dung beetles, they are not so numerous or speciate. The only major area of land left on the planet in which there are no scarabaeine dung beetles is Greenland.

The switch from general scavenging to dung is nicely manifested in dung beetle jaws. The ancestral jaws (secateurs are a good analogy) still retained by those scarabaeid beetles developing in 'hard' food (such as rotten wood, fungi, roots and dead leaves) are strong and designed for cutting and chomping relatively solid material. Dung beetle jaws are soft, fragile and sometimes membranous, adapted for soft, semi-liquid food. The mouth is covered with a flat protective shield, called an epistome, which also comes in useful when digging. Despite their large size and stout appearance, adult dung beetles eat only the tiniest of food particles, and their jaws are not able to take large chunks. This is very important when handling big shiny dung beetles; it's quite impressive (to the uninitiated onlooker) for the budding scatologist to pick up even the largest scarab or dumbledor, knowing that it will not try to take a bite out of your finger. Take care with any very large grubs though: they can nip.

The adults' jaws ('mandibles') are specially shaped for dung-feeding. The upper part (the incisor lobe) strains out fragments that are too large to be eaten, allowing only small particles (2–200 μm) through. Tiny though these may be, they are still chewed smaller before they are swallowed. The lower part of the mandible (the molar lobe) is an abrasive surface that grinds down much of its food to less than 1 μm (1/1,000 mm). This fits well with our knowledge of plant decomposition. Leaf litter, dead wood, fungi and humus need a lot of chewing, but as decomposition progresses (with the assistance of micro-organisms) it becomes more liquid; it is this liquid, rich in carbohydrates and proteins, which is most nutritious. Dung, by virtue of the digestive tract through which it has passed, is already rich in microbes and is watery, or at least moist, to benefit the nutrient absorption by its previous owner. Dung beetles take immediate advantage of that dung soup, without having to clutter their own bellies with tough, fibrous, bulky roughage. They leave that to their larvae.

What has clinched the excremental habitat for today's diverse and numerous dung beetle fauna was not just an early switch from broad scavenging in each lineage, nor adjustments to tooth and jaw, nor a taste for the soup of the day, but a propensity for nesting behaviour.

A BEETLE IN THE NEST IS WORTH TWO IN THE LEAF LITTER

The reason nesting behaviour becomes so important in dung beetle divergence is that if you are a simple organism like an insect, and you are going to make a nest, more or less under blind instinctive control, you have to commit to a particular way of doing it. It's no good landing on the forest floor, having a look around to see what materials are available, and then deciding what's the best way to gather them into some sort of lump for your offspring. This sort of random activity might work if there is plenty of food lying about, and there are plenty of eggs to be laid, but it will open any offspring, and hence the future of the race, to all sorts of competition, predation, parasitism and exposure to the elements. Much better to make a commitment to a particular type of detritus and do something in particular with it. This is the route down which proper dung beetles have trudged – rolling, scraping, carrying their fetid burdens.

Nesting behaviour is a tremendously successful trait. And it has paid off in spades for dung beetles. They may have committed themselves to a fragmented and transient foodstuff in the wider landscape, but they have gained by quite literally carving it up and doling it out among themselves. Tunnelling is seen as a primitive nesting strategy, the first type of rudimentary behaviour to have evolved back deep in prehistory. By choosing slightly different tunnelling techniques, different depths, different lengths and at different angles, the dung beetles have created extremely specialist niches for themselves and their offspring, allowing incredibly diverse faunas to develop, along with that bizarre array of ludicrous horns.

It's guesswork now, but it seems reasonable to assume that tunnelling started as a meagre scrape in the soil and the removal of a small morsel of dung to act as a larder. Once they were off, though, tunnellers sank to new depths, ever deeper and more complex, to avoid competition. Burrows can be shallowly inclined to nearly vertical; stashes of dung are taken down to the tunnel ends, or stuffed into off-corridor cells. In loose sandy soil, the walls may be lined with a smear of dung to reinforce them. With dung-rolling arising later, a new spate of ever so slightly different dung-burying techniques allowed a new era of divergence and speciation. A good overview of roller nesting ecology is given by Halffter *et al.* (2011). Further, faster rolling, in better straighter lines, using the sun, moon or stars, charted their course to the future. They've never looked back.

Back in the block of dung itself, of course, the dwellers have still managed to diversify. In northern temperate regions the dwellers (mainly the family Aphodiidae) dominate, even though any nesting behaviour is pretty primitive at best. This they can do because of better dung availability, and the fact that the dung does not dry out too quickly. They have avoided competition by evolving different sizes, specialising in different habitats or appearing at different times in the year.

There is still overlap with the nesters. One group of putative tunnellers (*Oniticellus*) live in dry-season forest elephant dung (where competition is arguably less intense), but instead of nesting underground, the female makes a clutch of brood balls in a cavity void shaped inside the dung heap, where she keeps guard over them well into their larval development. And some *Aphodius* dwellers are quite happy to leave the sanctuary of the dung to go on raiding expeditions down into the tunnels, becoming cuckoo parasites on a regular basis, or as opportunity arises.

WALKING WITH DINOSAUR DUNG?

True dung beetles first appear in the fossil record during the late Jurassic, around 140 MYA. No prizes for knowing that this was also the time of the dinosaurs. There is a general scientific agreement that ecosystems at that time were broadly similar to those of the modern era, the Cenozoic, which began with the extinction of the dinosaurs and much else besides, after the Chicxulub asteroid impact 65 MYA.

Although they don't feature much in dinosaur dioramas in museums around the world, there has also been a wide supposition that dung beetles had similar roles to those they act out today, just that they would have been shovelling sauropod scats, rather than road apples. This ain't necessarily so.

Today, whether produced by herbivore or omnivore, it is mainly mammal dung that is visited by dung beetles. Birds (the only living descendants of dinosaurs) and reptiles produce droppings that are not at all attractive to beetles. Of 29 species found in accumulations of bird droppings in Venezuela (Peck and Kukalova-Peck 1989), only one 'dung beetle', *Anaides fossulatus*, was found, but this is more regularly a decaying-organic-matter species, and is not usually attracted to dung. And there is a single report of *Canthidium ardens* dolefully pushing a chicken dropping in Panama.

In New Zealand, there was no history of large mammals until Polynesian settlers arrived only about 750 years ago. The plentiful dung beetles (only 15 native species though) had nevertheless evolved to become generalist scavengers. They may have occurred sporadically in the occasional moa dropping, but today they are mostly found on carrion and generally on rotting plant material (Stavert *et al.* 2014).

Likewise, reptile faeces are mostly ignored. Back in Panama (Young 1981a), only two scarabs (*Onthophagus sharpi* and *Canthon moniliatus*) turned up in traps baited with iguana and boa constrictor dung, but these were dismissed as having wide feeding niches anyway, and occurred in very low numbers compared to traps nearby loaded with mammal dung. Panamanian toad faeces (toad stools?) fared a little better, attracting five species, but still in low numbers. I've wasted many a happy hour poking about in goose droppings looking for dung beetles. Sadly, I found nothing, though the only modern UK record of the mainly Franco-Iberian *Onthophagus furcatus* was in a goose dropping at Kew. Being plant-eaters, goose faeces ('gaeces' has been suggested) are about as herbivore as they come, but this is still not quite right for most true dung beetles.

Going back to digestion basics, this is because birds combine their intestinal waste (what would become dung) with kidney filtrate (urine) in the cloaca storage organ at the end of the body, and eject them together in a wet chequered black and white dropping. The resulting excrement is high in ammonia and uric acid compounds, phosphates, carbonic acids and salts; dung beetles don't like these, which is why accumulations of bird droppings can amass into guano mountains, and are not recycled by the beetles. By extrapolation, dinosaur droppings must have been just like giant bird splashes (Arillo and Ortuno 2008).

There is some merit in the argument that the large droppings from huge herbivores such as *Diplodocus, Stegosaurus* and *Triceratops* could have had enough coarse plant material in them to produce dung boluses which would remain separate from the more liquid urinary waste, even if they were voided together. DNA analysis of various dung beetle lineages now suggests that many of the distinctive family and subfamily branches we know today started to diverge in the middle of the Cretaceous (120–130 MYA), and it has been suggested that it was a change of dinosaur diet from tough fibrous conifer gymnosperms to more nutritious flowering angiosperm plants that kick-started this radiation (Gunter *et al.*, 2016).

Certainly fossil dinosaur droppings (coprolites) are widely known, and a few of these show signs of having been burrowed into by dung-feeding creatures. Some of these ichnofossils (trace fossils, like footprints or larval galleries, but without the animal remains) have been named, but whether they are from fly maggots or beetle grubs remains conjecture. The earliest almost certain dung beetle ichnofossil is *Coprinisphaera*, from late Cretaceous (approximately 80 MYA) South America, showing fossilised dung beetle pupal chambers. One immediate problem is that the dung from which these rough balls were created is unidentified. Of course the problem here is that fossils are usually dated by their vertical position in the layered rock strata, and burrowing dung beetles tunnelling vertically down into the soil might end up, along with their dung balls, fossilised in a stratum from before they actually lived. A conundrum for sure.

However they might have first evolved, it was the rise of the mammals, after the disappearance of dinosaurs, and particularly the evolution of grass something like 60 MYA, that appears to have driven dung beetle diversification to the burgeoning genus and species numbers we see today. Now comes some rather speculative suggestions from dung beetle biologists. Having earlier pooh-poohed the idea of random gatherings of decaying material on the forest floor into a nest, this may very well be the way that nesting behaviour began. The argument goes that mammalian dung is little different from rich soil humus, and that some sort of precursor dung beetle was already gathering decaying leaf litter together into nesting clumps for its young. When mammals started to evolve and diverge, to fill the ecological voids left by the extinct dinosaurs and a whole host of other organisms, these beetles were ready and waiting, equipped with a rudimentary nesting instinct, and all set to exploit the dung that was dropping larger, harder and faster with every millennium.

ONCE A DUNG BEETLE, ALWAYS A DUNG BEETLE?

In a word – no. Dung beetles have a look about them: that thick-set, chunky form, strong shovelling legs, flat broad head, the fanned plate-like antennal club segments, and the fact that the males dress to the right (Medina *et al.*, 2013). Even to the untrained eye, they soon become familiar. So it is always a surprise to an entomologist that something quite obviously a dung beetle should be found doing something well away from dung. But there are plenty of examples.

Paraphytus aphodioides is a small shiny black 'dung' beetle that lives under the bark of large fallen trees in west Africa. Here, in the dank fruity fungoid decay, very like rich soil humus, the female gathers a ball of sawdust and broken wood particles and lays her egg in it. *Paraphytus* occurs with the wood-boring bess or passalid beetles (family Passalidae); these are closely related to stag beetles, and in the same lamelliform clubbed-antennae group as dung beetles themselves. There is some reason to suppose that *Paraphytus* uses ground-up chewings left by the bess beetles, and their larvae, mixed in with bess beetle frass; so they might, at a stretch, still be considered dung beetles (Cambefort and Walter 1985). The passalids live in semisocial family groups, adults and larvae together, so *Paraphytus* might also be considered a cuckoo-type parasite in their nests. Little else is known about either beetle's behaviour, but it would be fascinating to study whether the bess beetles react to the presence of interlopers.

Three Australian rollers in the endemic genus *Cephalodesmius* have lost the ability to roll dung; in fact they have lost their taste for dung completely. They have not reverted to scavenging, but moved on to leaf-cutting. Male-and-female pairs dig a burrow in the rainforest floor which they stuff with leaf material cut by the male. He doesn't have mandibles capable of biting, but uses the sharp edge of his front legs in conjunction with prongs on the front of his face. The female then chews the plant fragments into a paste-like synthetic dung; she shapes them into four or five balls, and lays her eggs in them. The brood balls are not large enough for each developing grub; instead the female continues to supply newly chewed leaf material as the young develop. By moving away from scarce and sporadic dung, *Cephalodesmius* is able to reach densities of 20,000–50,000 per hectare (Dalgleish and Elgar 2005). The Tasmanian grass grub, *Aphodius tasmaniae*, does a similar thing in southeast Australia, and New Zealand, and is sometimes reported as a minor nuisance in domestic lawns and golf courses. European dor beetles of the genus *Lethrus* have huge sharp mandibles with which they clip leaves. They then fill a subterranean burrow with a sausage-shaped mash of chewed leaf material for their brood. Large numbers can make unsightly holes in the herbage, and they are considered vineyard pests in Hungary.

Dor beetles (Geotrupidae) seem to be split into those species that utilise dung and those that choose leaf litter (like the Florida deepdigger in chapter 5), but the flightless Iberian dumbledor *Thorectes lusitanicus* seems to be stuck halfway. For breeding purposes it buries balls of sheep, goat, deer or rabbit dung and lays its eggs in them. But it will also bury acorns (Péréz-Ramos *et al.* 2007). The acorns of the Algerian oak, *Quercus canariensis*, and the cork oak, *Quercus suber*, drop in autumn, and the beetles bury them a few centimetres into the soil. They are then able to break open the tough woody seed and start eating it, eventually hollowing it out completely and sometimes overwintering inside the acorn shell. Between a half and one-third of buried acorns are eaten, but this means that a large number are left intact, presumably lost and forgotten, and could potentially germinate. Since most fallen acorns are cleared away and eaten by foraging birds and animals, the trees benefit from the unusual behaviour of the beetles. Meanwhile in South Africa, *Ceratocaryum argenteum*, a rush-like plant endemic to the Cape Province, drops seeds about 15 mm across, which mimic antelope dung by smelling distinctly of faeces (Midgley *et al.* 2015). The roller *Epirinus flagellatus* is happy to roll the seeds away and bury them, but apparently derives no benefit. If it uses real dung it remains with the brood ball, but buried *Ceratocaryum* seeds are always abandoned. Maybe it only realises it has been duped when it has done all the hard work.

The distinctions between dung, rich humus and leaf-litter are often blurred. The putatively obvious dung beetle *Aphodius distinctus* is a European species introduced into North America, where, like many novel immigrants, it can be extremely common – 1,097 specimens in a cow pat not 2 hours old (Floate 2011), another of those famous dung beetle statistics. But come spring, when farmers are muck-spreading, the fields can be ablaze with clouds of them flying everywhere over the turf. The larvae are more general detritivores than many dung beetles and live quite happily in the enriched humus rather than fresh dung, with densities of $90/m^2$ reported; they can be reared simply by keeping them in jars of the soil. It's actually not uncommon to find various *Aphodius* species in putrescent fungi, under the well-rotted fungoid bark of old logs, or in garden compost bins. They are perhaps, living on the gentle end of a sliding scale along a spectrum of putridity, but right up there with the most fragrant is carrion.

There are plenty of examples of dung beetles occasionally using vertebrate carrion instead of dung. In the genera *Coprophaenus, Deltochilum* and *Canthon*, some species are dung-feeders, many others are carrion specialists. Choosing a putrescent corpse rather than a fragrant pat, morsels of decaying flesh are ripped off and shaped into similar brood balls, then rolled away or buried in tunnels. In Australia, a peculiar continent where harsh ecological constraints have had some bizarre effects on the evolution of many animals, the native *Onthophagus consentaneus* did not ever feed on marsupial dung, but probably subsisted entirely on carrion and decaying fungi. It has benefited from modern human impacts there, by utilising abudant road kill. In Oklahoma, specimens of the small *Canthon imitator* (formerly *C. laevis*) were seen to roll up and make off with portions of dead tadpole instead of dung (Bragg 1957), and I recently saw a video of a large dung beetle rolling a much larger dead lizard over the ground: this is ambition beyond the call of duty.

One of the oddest onward adaptations away from dung is shown in the two known species of neotropical *Zonocopris*. These small

(2.5–5.0 mm), elegant, curvaceous dung beetles feed, as adults, on snail slime, appearing to spend their entire lives on living giant *Bulimus* and *Magalobulimus* snails. They do not harm the snails, and although mating has been observed in captivity, no egg-laying has ever been observed (Vaz-de-Mellow 2007). Their larval food remains a mystery, though snail faeces or dead snail hosts have been suggested as possibilities.

They get weirder. One of nearly 90 species of the New World roller genus *Deltochilum* feeds on millipedes, which it first hunts out and kills (Larsen *et al.* 2009). *D. valgum* is attracted to the millipedes in the rainforests of South and Central America, especially injured specimens, which give off a malodorous defensive secretion that vertebrate predators at least find repulsive. To a dung beetle, though, this is as a scent of nectar and honey. *D. valgum* uses its strongly curved hind legs (much more bowed than others in the genus) to grip a millipede, often much bigger than itself. It then pushes in, to disarticulate or decapitate its prey using its narrow head, which is sharply pronged along its razor-sharp front edge. Not exactly a head-butt, more of a head-stab. Adult beetles sometimes fight over their prizes. The millipede prey is not buried, but is carted off using the beetle's curved hind leg, gripping it against its pointy upturned tail tip (not found in others in the genus), and hidden in leaf litter where an egg is laid in it. *D. valgum* only ever feeds on millipedes, but a few others in the genus are sometimes attracted to millipede scents, suggesting that this behaviour is on the cusp, in terms of evolutionary divergence. Elsewhere in South America, the small roller *Canthon virens* similarly attacks large leaf-cutter ants and buries them as larval food (Hertel and Colli 1998).

My favourite, though, without question, is the aptly named *Eocorythoderus incredibilis* a tiny (3 mm) shining red-brown scarabaeid, elegantly shaped with a narrow waist, and wonderfully described as panduriform – violin-shaped, like a pandura (or pandora), an archaic three-stringed lute (Maruyama 2012). It only occurs in the subterranean fungus gardens (on which it feeds) of the termite *Macrotermes gilvus*, deep in the Cambodian forests around Angkor Wat. More than 20 other genera of 'dung' beetles are found solely in the nests of ants and termites. *Eocorythoderus* is flightless, with only tiny eyes, and relies on the termites to move it about using a carrying handle node at the base of its wingcases. They grip this in their jaws and transport it about from room to room in their long-lived nests, just as they do their own nymphs. In my book this is officially the cutest insect in the world.

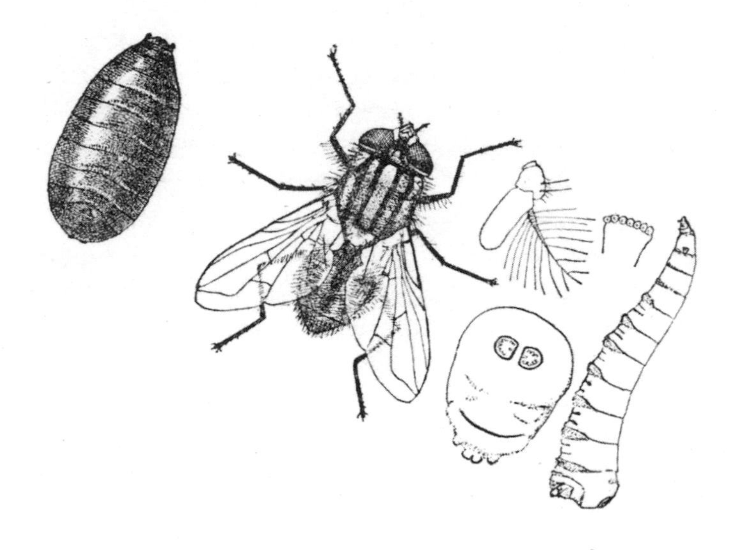

CHAPTER 7

A CLOSER LOOK – WHO LIVES IN DUNG?

THE ANCIENTS KNEW, only too well, who lived in dung. And they celebrated these creatures with idolatrous zeal. Obviously, my favourite exhibit at the British Museum is the 1.5 m long, 1 m high, giant stone sculpture of a sacred scarab. It is beautiful. It was created around 332–330 BC, during the Egyptian Ptolemaic period. This was something like 1,800–2,300 years after scarabs were truly held sacred, and was made during a period of sentimental nostalgia for the good old days. Although, perhaps, a pastiche, it's an awe-inspiring piece of art – figurative yet symbolic, mundane yet iconic. It draws from a familiarity with the natural world of excrement that might surprise today's prudish, urban, urbane, reader. At the height of scarabmania, during the Early Middle Kingdom, around 2000 BC, scarab amulets, necklaces and brooches were by far the most popular items of domestic jewellery; many thousands of them have been unearthed in archaeological digs and there was a Mediterranean-wide industry in manufacturing them, from Sardinia to the Levant.

Though they represent the same type of creature, these models are not all the same. It seems clear that various artisans, in different localities, throughout the centuries, have used a variety of prototypes on which to base their works. Such is the number of well-preserved scarab artefacts that one entomologist (Klausnitzer 1981) has tried to identify which genera might be represented: *Scarabaeus, Catharsius, Gymnopleurus, Copris* and *Hypselogenia,* he suggests, and his side-by-side comparisons of beetle outlines and carved effigies are pretty plausible. I don't think there is any suggestion that ancient scarab carvers knew much about dung beetle classification, or even that there might have been different species, but the fact that we can look at their art, 4,000 years later, and identify the different insects shows, surely, that they had an intimate knowledge of the beetles, that they had seen them, held them in their hands, and examined them carefully enough to reproduce them so accurately in their carvings and quartz ceramic faience modelling.

Fig. 29 Some ancient Egyptian scarab representations were more fanciful than others, but their sheer numbers and diversity speak of the reverence directed at dung beetles.

In this chapter I'd like to suggest that not enough people get to appreciate or handle dung beetles; or dung flies, or the many other dung critters for that matter. Which is a shame. I've been fascinated by dung and its denizens since I was a boy. The very first entomological survey I ever wrote up, in my awkward 17-year-old prose, using my best schoolroom handwriting in a large ruled hardback exercise book, and with a scattering of amateurish sketch drawings pasted in, was pompously entitled 'The South Heighton *Aphodius*, being an account of those species in the genus *Aphodius*, in the family Scarabaeidae... etc, etc... found in the parish of South Heighton and parts of Newhaven, Tarring Neville, Beddingham... etc, etc... between 1970 and 1975.' Just so there was no doubt, I dated it MCMLXXV. I think I was reading too much Victorian literature at the time. It's tucked away in a corner of my bookshelves now, to remain obscure and unread for a couple of hundred years, when it might be unearthed as a curious historical artefact. My interest in decaying organic matter has stayed with me, and this chapter is, on a more personal note, an exploration of my journey through excreta. Here I'll be giving details of some of my favourite dung animals, and how to go about finding and observing them. But first a necessary health warning.

NOW WASH YOUR HANDS

Dung is, as first noted in chapter 1, largely made up of bacteria; and these are mostly the kind of bacteria that you don't want to end up on your food. So, rule number 1: if you're going to become a scatologist, always wash your hands before eating.

Some years ago I was taken to task after writing what I thought was an amusing article about collecting insects (Jones 1986). In it I related the specialist wet-finger technique for collecting small beetles. This is perfect if you see a tiny beetle running across a hard surface such as a fence post or a log. You simply lick the end of your finger, dab it onto the beetle, lift it up by the adhesion of your spittle on the glossy dome of its wing cases, place a glass collecting tube over the wriggling insect stuck on your finger, and give your hand a smart tap; this dislodges the insect down into the tube. Voilà. I went on to make a joke about how the flavour of one's finger changes through the day, especially if you have been out dung-beetling. Sadly, my cavalier attitude to personal hygiene was found to be rather tiresome, and someone was obliged to point out that this is just the way to pick up nasty intestinal complaints. Possibly even fatal diseases. They said so in a letter to the editor.

Proper books on dung beetles will insist that disposable rubber gloves should be worn at all times. And this does add something of a formal surgical nature to the dissection of the pat. Many students of dung avoid handling it altogether by using the immersion technique for finding dung beetles – simply put the entire dropping into a large bucket of water. As it disintegrates, the insects within float to the surface, where they can be seen waving their legs about. They can then be hooked out. This doesn't always work though; *Aphodius granarius* was always regarded as being very scarce in Finland, which is, as everyone knows, an international centre for dung beetle research. This is because it does not rise to the surface of the water. Manual searching of pats, however, revealed that it is in fact widespread throughout Scandinavia, and frequently found in the UK. The Finns need to abandon their buckets and sharpen their trowels.

You don't actually need to get down there and in it, to find dung insects. You can observe closely, or at a distance, and watch the comings and goings of endless flies and beetles. Take the advice of a hardened dung-beetler, Eric Matthews, previously of the University of Puerto Rico. He wrote a seminal 18-page paper (Matthews 1963) on the rolling behaviour of the tumblebug, *Canthon pilularius*, and he clearly states his modus operandi:

> *The methods used consisted of sitting down and watching the activities of the scarabs.... The author always camped in the immediate area of observation so that he could be present for every phase of activity from beginning to end. Often days were spent in the same spot.... No special techniques were used and no experimentation was attempted.*

In other words, he wrote about what he did on his holidays. What a lark, eh? At the other extreme, I am hugely impressed by the energy and verve of another US entomologist, Leland Ossian Howard, whose 1900 paper is a monument to the study of insects associated with human dung (Howard 1900). His was not just a personal exploration of a few sample stools; it was an intense and detailed forensic analysis of large army camp latrines. This was supreme dedication, and his paper is beautifully illustrated with engravings of all the most important flies, and their maggots. For more inspiration, I can thoroughly recommend the excellent general introductions to northern hemisphere temperate dung faunas by Putman (1983), Skidmore (1991) and Floate (2011).

At the gentler end of one's leisure activities, dung study can be contemplated during a country walk, picnic or safari. You can simply sit and watch the farm animals, or the wildlife, and what they are depositing. If you are lucky enough to be in the vicinity of fresh elephant dung in the spring or summer savannah, you may be inundated by thousands of them. Anywhere south of Bordeaux you might see rollers at work, and be inspired just as were the ancient Egyptians. Just rest easy and watch them go about their business. If you do decide to explore deeper, I recommend a narrow garden trowel, or a stout penknife, but if this is not to hand it is easy enough to tip over the dropping with a stick, to see what is sheltering beneath. So it was when I found myself beside some unusually large spoor (dog I guessed, rather than bear) in a disused Florida orange grove back in 1991. Flicking it over with a dead twig, I uncovered the exquisitely viridescent rainbow scarab, *Phanaeus vindex*, looking as if it had been shaped from beaten gold and copper, and sporting, in the male, a huge back-curved head horn. Wonderful.

For larger cow and horse droppings, start at the edge, prising apart any natural fissures, and looking into the grass roots under the dung. Small excavations in the dung may show where a diminutive dung beetle is hollowing out a morsel. Sometimes just the tail end and the back legs are visible. A pile of earth beside or under the dung will be the spoil from a tunneller. Unless you are prepared for serious digging, it may be enough to expose the burrow entrance and wait; eventually the beetle may come back up to the surface. Serious entomologists will stick something like a straw into the hole, then dig down until a deeper section of the tunnel is exposed and the process can be repeated. Be warned, this can go on for several metres.

Dung heaps are different from dung, and not a few entomologists insist on the term manure heaps by way of distinguishing them. They have straw, sawdust, hay or whatever other stable, hutch or barn bedding is mixed in with them. Today they are relatively transient piles, easily moved around by JCB and tractor, but in times past they were long-lived, added to, and taken away from, piecemeal, as the farm animals provided and the farmer needed. Over many years they can mature and acquire an important diverse community. They might have fewer strictly dung beetles, but are still rich in decaying organic matter fauna, and they can get quite ripe, in a fermenting, mouldering kind of way. I'm very much a get-stuck-in-there naturalist, but I well remember Earnest Lewis, a coleopterist friend of my father's, sitting down in smart three-piece suit and tie, and polished leather shoes, to dissect the edge of a manure heap on a short country walk when he came to visit us. He took a small trowel and a plastic sheet and could have gone on to have cocktails with the squire later in the day. I used my bare hands and probably needed a bath before my Mum let me back in the house. So, please, when you're done, do wash your hands.

THE ENGLISH SCARAB – NOT SO SACRED

We don't get the roller scarabs, sacred or otherwise, in the UK, and for this I am sad. Our dung fauna is impoverished compared to a mainland Europe only a few miles away over the water, and it is also outperformed by Scandinavia. This is a blunt geohistorical fact, an unfortunate consequence of retreating glaciers after the last ice age and the flooding of Doggerland as it sank between the North Sea and the English Channel, effectively cutting us off from colonists moving north as global temperatures rose 15 millennia ago. Nevertheless there are some real gems amongst the 400 or so dung-associated invertebrates here. And travelling to foreign climes, where the dung is remarkably similar to that in Britain, is a real adventure to be seized with relish.

The nearest we get to a proper scarab in the UK is in the large glossy domed form of *Copris lunaris*. At one time it was grandly called the English scarab, which is a bit odd because this exceedingly handsome beetle no longer occurs in England, or anywhere in the British Isles for that matter, unless you count Jersey and Guernsey, and most people don't. The name was coined, slightly after the fact, it has to be said, when conservation organisations were writing biodiversity action plans for various charismatic endangered species during the 1990s. Everything needed an English name, otherwise journalists and politicians, let alone the general public, wouldn't have a clue. By then it was already considered extinct in the UK. It was always very rare, and was last definitely seen here when a specimen flew into the entrance hall of the Juniper Hall Field Station near Box Hill, on the North Downs in Surrey, at 10.30 p.m., on 27 May 1955. This area of ancient grazed chalk downland would have been perfect.

I grew up on the South Downs, in East Sussex, and this was always a beetle I half fancied I might find. In my dreams. Except my dreams just might have come true, if I'd been lucky enough. In 1994 Peter Hodge was given a specimen of *Copris* found in a glass-topped display case of insects, made at some time past by a local naturalist, H.L. Gray, and later donated by his widow to Lancing College, in West Sussex. The tantalising handwritten data label stated 'Lancing Ring, 19.9.60'. The first thought was that this might refer to 1860, when this species was known from nearby Shoreham, but the beetle had

Fig. 30 A unicorn amongst beetles, *Copris lunaris*, marvellously structured, enchantingly elegant, mythically rare. An affectionate engraving from Michelet (1875).

been mounted on the back of card cut from some printed material; from the typography, and other beetle mounts in the collection, it could be dated directly to 1960 (Hodge 1995). I had been kicking around these exact same chalk downs a decade and a half later, but never saw it. I tried my darnedest though.

At least it is pretty widespread in France, and I got a great thrill finding it in the meadowland attached to a sun-bleached holiday gîte in southern Burgundy a few years back. I did not need to self-bait for it. It's a magnificent creature: the male has a long back-curved horn on its head (a feisty tunneller, no doubt), the female a slightly smaller one. Its body is sturdy, elegantly rectangulo-subovate, gently ribbed along its wing-cases, delicately dimpled across the thorax, entirely a lustrous black, looking for all the world as if it were sculpted from obsidian. I'm going to commission my own scarab amulet, based on this mythically rare beast – a unicorn amongst beetles.

Another unicorn, but one that does turn up in Britain occasionally, is the almost globular *Odonteus armiger*. Whether it is a dung beetle or not is still being teased out. Current consensus is that it feeds on subterranean fungi. It has been found burrowing down 30–40 cm to get at them. There was a notion, not very convincing, that it is associated with rabbit warrens, the fungi somehow benefiting from the buried crottels, implying that the beetle was an indirect coprovore. Old beetle monographs state things like 'In dung; generally taken on the wing.' They also like to relate things like: 'Mr Mason's specimen [Croydon, 1880s] is one of the most recent instances of its capture in Britain; seeing a beetle flying past, he knocked it down with his stick to see what it was, and found it to be this very rare species' (Fowler 1890). I've never seen one alive, but occasionally I hear of a sighting, and am lost to a moment of quiet reverie. And I wonder if, in my dotage, I will be as accurate with my cane, as I potter about in the grounds of the care home.

These are, admittedly, unlikely finds, but the common dor beetles (Geotrupidae) are just as impressive. When my family first moved to Newhaven, in Sussex, after my father's office relocated to nearby Lewes in 1965, we had the South Downs on our doorstep. It wasn't long before we were off exploring the rolling hills around South Heighton, Tarring Neville and Firle. Many of these were grazing meadows and although it would take a few more years before I really got stuck into dung beetles, I well remember Dad and me finding scores of *Geotrupes* on one evening stroll to nearby Bishopstone.

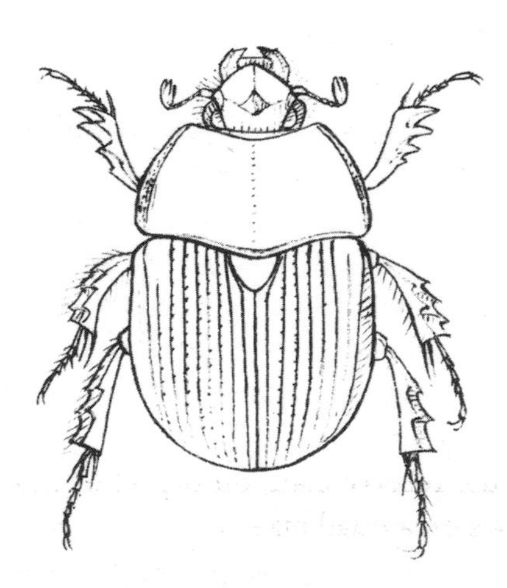

Fig. 31 A dumbledor, or dor beetle, *Geotrupes*, a picture of the robust earth mover.

They were buzzing about, low, bumbling, over the grass, and we didn't need a net to catch them. I was astonished at their power, as one clawed its dumpy clockwork way out of my cupped hands, pushing its smooth blunt head forwards, as it heaved with its broad toothed legs. When it took to the air it was like a miniature helicopter taking off from my palm, and I can almost feel the memory of that downdraft on my skin now. And somewhere at the back of my brain is the gentle buzzing note they make as they fly past; the sound of a summer evening in a childhood of blessed memory.

The name 'dor' is derived, according to my dictionaries, from the sound of the insects flying, and although these books often suggest bumblebees (or 'humblebees'), any entomologist worth their salt will tell you that bees hardly buzz when they're in flight. 'Dor' definitely applies to these giant dung beetles. I'm not sure if it was dung beetles or bumblebees that inspired J.K. Rowling to call Harry Potter's headmaster Dumbledore (using the later alternative spelling), but when I snatched one out of the air on the banks of the Wolfgangsee near Salzburg in 2013, my Austrian in-laws were immediately familiar with the professor's name. They remained politely enthralled as I enthusiastically introduced them to the ecological concepts behind the natural recycling of decaying organic matter.

AN INSECT TO BE PROUD OF

Although it sounds a bit like a bottle of craft ale or a cross-Channel ferry, the pride of Kent is actually a beetle. *Emus hirtus* is an enormous furry rove beetle; again it's very rare, and I've never seen one. It is not, however, extinct, and has been reported regularly, if exceedingly sporadically, in the Elmley Marshes National Nature Reserve, on the Isle of Sheppey. It first turned up there in the gents' toilets near the car park in 1997, but puts in an occasional appearance for the reserve wardens or visiting naturalists. A flying specimen was also reputedly caught by one of the reserve staff, on his bare chest, as he motored round the meadows on a quad bike.

The rove beetles are an exceptionally successful group of insects. They have very short wing-cases, under which tightly folded membranous flight wings are protected. This gives them the double advantage of having an extremely flexible body form to wriggle into tight spaces, but still retaining the ability to take to the air. Plenty live in dung, from the flat slow dung-feeders such as *Oxytelus* and *Anotylus*, to the predatory *Philonthus* and parasitoid *Aleochara*. *Emus* is a fierce predator, a fast and efficient killing machine that will tackle flies, maggots, grubs, dung beetles; indeed whatever it can get its teeth into. It only ever seems to appear at fresh cow dung, and coleopterists visiting Elmley can be identified by their insistence on closely following the cow herds around (or being closely followed by the cows according to RSPB naturalist Rosie Earwaker), in the hope of spotting *Emus* at a steaming pile of excrement only minutes old (Telfer *et al.* 2004).

Unlike the smooth and shiny scarabs, *Emus* is covered with grey and gold fur, and when it flies it looks just like a manic bumblebee. Accounts of its activities at the pat vary. The late A.M. Massee (a respected and knowledgeable entomologist, but also a notorious raconteur) reported that he had seen *Emus* fly down to a fresh liquid cow pat, dive in and emerge a short time later, its coat of luxurious golden hairs remarkably unsullied. This is slightly at odds with the advice given by Föreningen SydOstEntomologerna (South-east Sweden Entomological Society) which suggests that to catch *Emus* 'smack the beetle with a firm hand down into the smeary dung, find it by digging and put it in your tube with some absorbent tissue', implying that the beetle becomes mired. I'm sure they washed their hands afterwards.

I suspect that *Emus* avoids clogging its fur coat by pushing under the dung, amongst the grass root thatch, rather than trying to swim in the gloop. This is exactly what the slightly smaller, and slightly more demure, *Ontholestes tessellatus* does. It's not so large as *Emus*, but is still attractively decorated with a variegated pattern of glistening metallic brownish hairs. It too seems to avoid becoming sullied, despite a similar predilection for glistening, ultra-fresh dung. *Ontholestes* is always intent on catching flies and is hyperactive on the fresh dropping, flicking backwards and forwards with a frenetic, jerky gait.

Other show-off rove beetles to look out for, disporting natty black and red tones this time, are *Philonthus spinipes*, a recent spiny-legged arrival into Europe from the Far East, and *Platydracus stercorarius*, the generic name of which aptly translates as 'flat dragon'. Again, they are fast, agile hunters, gone in a twinkling.

FLIES – THE GOOD, THE BAD AND THE BUGLY

Finding the yellow dung fly, *Scathophaga stercoraria*, wins nobody any prizes. This is a common, verily ubiquitous, insect and, just like me and the other keen first-year university students in Ashdown Forest, all you need do is sit close to a fresh animal dropping, and they come zooming in. The males, with their fuzzy yellow plumage, are very distinctive, but amongst them are the olive grey females. To be honest, it's easier to find the females sitting about on the herbage; here they are hunters, attacking and eating other small flies. As soon as they arrive at the pat, though, they become embroiled in a mêlée of over-amorous males. After the pat has dried a little, and developed a tough fibrous rind, the adult dung flies will have moved on, but lift up the lid of the pat, and long, white, narrowly conical maggots squirming in the ooze at least show that the flies have done their job here.

It's when the dung starts to dry out that one of the most charismatic of flies may deign to visit. The hornet robber fly, *Asilus crabroniformis*, is northern Europe's biggest fly, and although its black

Fig. 32 *Asilus crabroniformis*, disarmingly illustrated by Shaw (1806), but it is the mouth end that does the damage, at least to flying dung beetles, not the pointed tail.

and yellow colour scheme may evoke a fear of waspish stinging, it is the mouth end that delivers the danger. This ferocious predator sits atop a dry pat, darting up rapidly to snatch flying insects out of the air. It is quite capable of wrestling a dor beetle and coming out on top. It has strong bristly legs, which it uses a bit like a grapple, and sharp piercing mouthparts with which it skewers its victim.

The family move to Newhaven soon brought this large insect to my attention, and I even have a vague memory of it appearing in the garden once or twice – though not on dung. The last time I saw it was in Poverty Bottom, one of the softly rounded glacier-hewn valleys through the South Downs behind Denton, where I sat one fine summer day of the school holidays, watching them taunt dragonflies and eat the occasional blow fly. One of our neighbours walked past, with his dog, and after watching me stalk a cow pat and swat my net over the top of it, enquired politely what I intended to do with the dung now that I had caught it. I hope I was able to make his eyes bulge with wonder, or fear, as I showed him the monstrous creature in the folds of the net.

For most of the other dung-feeding flies, I expect people narrow their eyes in disgust and loathing. The most obvious ones on the freshly dropped stool are greenbottles (*Lucilia*), bluebottles (*Cailiphora*) and mottled dung flies (*Polietes*). They can form a raucous bristling mass over the dung, taking to the air in deafening buzzing clouds if disturbed. Looking at a dropping alive with flies, it always amazes me that the links between excrement and fly-spread diseases took so long to be recognised. Mind you, they are still misunderstood and misrepresented today.

None of the flies mentioned above, nor for that matter the common yellow dung fly, *Scathophaga stercoraria*, or a whole host of others bombing or bobbing about on the pat, ever cause any real harm. They do not come indoors, are not attracted to human food and are not implicated in the spread of human diseases. They might fly irritatingly about your head (face flies, *Musca autumnalis*), or try to bite you (stable flies, *Stomoxys calcitrans*) when you're off walking through the countryside, but that's another matter. The only dung-breeding flies which do come indoors are the appropriately named house fly (*Musca domestica*) and lesser house fly (*Fannia canicularis*). *Fannia* is the small pale grey fly you often see flying in erratic zigzags beneath the hanging light, then landing adroitly upside down on the bottom of the light bulb. *Fannia* will not, though, fly down to land on your food. It is simply marking out a three-dimensional aerial territory in which to meet a potential mate; away from your kitchen these flies zigzag about under spreading branches or in the shade of an overhanging shrub. You will never see one on the dinner plate.

This, unfortunately, cannot be said for the house fly. When bacterial and microbial diseases first began to be understood in the late 19th century, it was the house fly which became the target of insect plague rhetoric. It wasn't long before more than 100 types of bacteria, virus and protozoa had been found hitching a lift on house flies, and a single house fly was found carrying 6.6 million bacteria (Hewitt 1914). Along with mosquitoes (which thankfully do not breed in dung), the house fly became public enemy number one; it was the insect menace to which book and poster campaigns were targeted. And all because it bred in filth.

Oddly, you don't find many house flies buzzing around in grazing meadows. Out in the wild they tend to be rather secretive. They also prefer manure heaps to fresh pats. This has had the strange effect of making them much less of house flies in the 21st century. In Britain, at least, they are no longer house flies, but farm flies. In 17 years of living in my current house in south-east London, I have found a house fly indoors just the once. I blamed the over-aromatic state of the guinea-pig's hutch, just outside the back door, for this. At other times there just is not the manure lying around any more. Along with the lesser earwig, house flies have declined in towns and cities, following the decline of horse-drawn transport. But whenever we go on holiday to rural France, we spend hours each day patrolling with the flip-flop of doom, leaving a trail of swatted corpses behind us.

House flies lay batches of about 150 eggs at a time, with half a dozen batches over a fly's lifetime of a week or a month. The maggots are fully fed in 3–30 days depending on the ambient temperature. A week or so later and the new adults have emerged. In temperate Eurasia and North America there can be a staggered range of 10–12 generations a year, but since they overlap the flies are constantly on the wing. In the tropics the fly is continually brooded and the flies can easily reach pest proportions. In Australia the house fly is replaced by the bush fly, *Musca vetustissima*, and if we thought we knew pest proportions in the Old World, European colonists to the southern hemisphere were totally unprepared for the multitudes on a biblical scale presented by this diminutive insect. The bush fly gets more prominent treatment in chapter 10.

Back at the manure heap, the suppurating fermentation is attracting lots of flies, and one of the most dramatic is the drone fly, *Eristalis tenax*. This large brown and orange hover fly is so named for its close resemblance to the male (drone) honeybee. It lays its eggs in nearby ditches, where the dank liquid outflow from the manure, or seeping spillage from the slurry lagoon, mixes with water draining from the fields, and also in streams where raw sewage is still discharged. The large pale maggot has a long 'rat tail', the telescopic breathing tube with which it takes air from the surface.[1]

The ancients understood a little about scarabs, but were not quite so up on fly ecology. The drone fly is the *bugonia* or oxen-born bee of myth, fable and the Bible. Until well into the Middle Ages it was authoritatively reported that a swarm of bees would appear, by spontaneous generation, from the body of a cow, especially one which had been bludgeoned to death, and had its nose and throat blocked to stop its soul leaking out. Feral honeybees will nest in hollow trees and rock crevices, but not in the rank putrescence of a rancid corpse. To the drone flies, though, liquefying carrion or draining faecal sludge are close enough decaying organic matter for either to suffice.

Samson, strong-man of the Israelites, was similarly mistaken when, wandering through the desert, he came across the dead body of a lion, buzzing with insects. He later offered a riddle to some truculent in-laws (uncircumcised Philistines): 'Out of the eater came forth meat, and out of the strong came forth sweetness' (Judges 14:14). His implication was that something edible ('meat' in an archaic sense) came from the predator, and that the strong lion was now a source of honey. Of course these were drone flies, nothing to do with honeybees – the mistake continues to this day, much to the frustration of picture editors worldwide. The myth of the oxen-born bee had been sorted out by the 19th century (Buckton 1895), but this

[1] *Eristalis* really does like the most rancid material. Geoff Hancock was pleased as punch when he found himself ankle deep in a moist tapir latrine in Costa Rica, and took the opportunity to collect and breed out several hover fly larvae, including that of *Eristalis alleni*, the immature stages of which had never been seen before.

hasn't stopped sugar manufacturers Tate & Lyle using the quote, along with a dead lion and a swarm of 'bees', on tins of their famous golden syrup. Go and visit the factory at West Silvertown, in the old London docklands, and you'll see a representation of the tin, the size of a transit van, projecting from the corner of the building. Bonkers and brilliant.

I realise we've ambled quite a distance from rural grazing meadows to industrial North Woolwich, so in an attempt to reboard the train of thought, I'm going to retrace my steps back to Poverty Bottom and the South Downs near Denton. It was somewhere around there, back in the late 1960s, that I first came face to face with the noon fly. Apparently *Mesembrina meridiana*, gets its 12 o'clock meridian name from its habit of basking in the full sun at the height of the day, and that's just about right. If dors are the symbols of dusk, then this huge, black, shining fly is the embodiment of the midday sun. Here it sits, in the crook of a stunted hawthorn bush, or on a tree trunk, soaking up the warmth, darting off rapidly if disturbed, but often returning a few moments later to its perch. It hardly ever visits dung; it doesn't need to.

In contrast to the profligacy of a house fly's many hundreds of eggs, or even the many scores laid by the yellow dung fly, *Mesembrina* may only lay five or six eggs, one at a time, a day or two apart, in the freshest dung it can find. The large egg has a spine-like process at one end, possibly to give it buoyancy in the liquid medium. The fly's frugal egg-laying policy echoes the slow but careful brood strategy of the large roller scarabs, but unlike them *Mesembrina* offers no maternal care to her young. The single egg deposit means a bigger egg, and a bigger hatchling maggot; and although this one will eat the dung itself, if there is nothing else around, it much prefers to eat other larvae. It is a formidable predator. Laying more than one egg per pat would be a counterproductive effort, as one would more than likely eat the other. By getting in quick, the female noon fly is trying to ensure that her single offspring will be top maggot, the biggest fish in the pond. Consequently, with only one egg to lay each day, *Mesembrina* can afford to sun herself leisurely most of the time. Taking a siesta.

There is no such luxury for the ant flies. The Sepsidae have nothing to do with ants, of course, but vaguely resemble them by virtue of their small size, shiny, black, narrow-waisted bodies and large bobble heads. Their most distinctive feature is a small black spot at the end of their wings. These are all the more prominent due to the flies' habit of slowly waving their wings about in a deranged semaphore. Sometimes both wings sweep together, sometimes they alternate, sometimes they gyrate in what looks like completely asynchronous chaos.

The purpose of the wing-waving is still unclear. In other flies such activity is closely associated with mating, and forms part of a complex who-are-you? mate recognition dance. But sepsids dance even when there is only one sex (males) about. Sometimes there are clouds of them, with aggregations of 30,000–50,000 recorded. These may be mass emergences of adults following hibernation. All-male clouds suggest lekking, a behaviour where males congregate in one spot to attract females. The fact that these congregations have a distinctive smell (at least to the human nose) suggests there may be some female-attracting pheromone scent involved. When I sat down in a grazing meadow near Hastings to eat my packed lunch one rather dull April day in 1976, I do not believe it was the odour of my ham sandwiches, or flask of tea, which brought out the cloud of many hundreds of wing-waggling sepsids from the grass tussock I was leaning against. Within minutes they were all over my backpack, crawling up my legs, getting in my hair. It was all very peculiar. I think I was just there in the right place at the right time to see this strange phenomenon. Fresh dung can attract huge numbers, although they have to be careful of the larger and much more dangerous yellow dung flies, which could quite easily mistake a sepsid for a tasty snack.

THE NOT QUITE SO SCENIC ROUTE

The horse bot fly maggot travels to the dung by a convoluted and circuitous route. It arrives already packaged in the dung, when it is dropped by the horse, having been living quite happily inside the animal's digestive tract for the last few months. It's tough exterior is completely immune to the plant-digesting processes going on in there. *Gasterophilus intestinalis* (that's Graeco-Latin for 'intestinal stomach-lover') is well named. It started out as a narrow white egg, one of up to 1,000 glued individually onto the hairs on the horse's flanks or front legs by the female fly, using a neatly upturned egg-laying tube at the tip of her abdomen. As the horse licks itself, the eggs are stimulated to hatch immediately and the tiny larvae burrow into the mucosal membranes of the poor animal's tongue and mouth. After about a month they allow themselves to be swallowed, and now half burrow to the lining of the horse's stomach using a series of hooked barbs around the mouthparts and skirts of backward-pointing bristles around the body. For the next 9–12 months they chew away at the stomach membranes, and with clusters of many hundreds of them in some victims, they can cause the hosts quite some distress, what with ulceration, possible infections and interference with normal digestion.

Eventually, when fully grown, the now large (1.5–2.0 cm), fat, dirty cream, bristly larva relinquishes its hold on the stomach wall and is passed down with the faeces, through the intestines to exit the horse and land in the field. Here it pushes out and burrows down into the soil, grass thatch or dried manure to harden into a dark spiky pupa. The adult fly emerges a few weeks later in late summer or autumn.

When I uncovered a freshly deposited maggot in a mound of horse droppings in a field between Fairwarp and Duddleswell, north of the equally unlikely sounding Uckfield, in East Sussex, in August 1974, I could not for the life of me imagine what it might be. By then I reckoned myself quite an expert on insects, and could tell it was some sort of giant fly larva, but its identity perplexed me. So I did what any sensible person would do – I took it home to rear it. Of course the secret to rearing unknown larvae is to take as much of the foodstuff too, so I stowed it in my now empty sandwich box with plenty of fresh horse dung. Simple.

I can't quite remember whether my Mum had anything to say about the matter. The exact details of our exchange evade me now. Thankfully it pupated within 24 hours and I was able to transfer it to a jam-jar of less pungent, therefore less complaint-worthy, garden soil. A fine orange and brown furry fly emerged later that year. I had never seen anything like it. I wonder if I will ever see it again. By 1974 this was already a scarce insect because it was already being widely persecuted, on the way to being eradicated. That's the trouble with agricultural pests, especially ones causing grievous bodily harm to horses; they somehow don't qualify for conservation status. Instead, they attract the attention of pharmaceutical companies, trying to poison them. The flanks of horses are now drenched with insecticide to kill the eggs. More chemicals are given as oral gels, liquids or feed additives to kill the maggots living in the gut. In Britain, at least, where there is no wild horse-like animal to act as a natural reservoir, it does not take much to eliminate the horse bot. It may still linger where feral horses roam, chemical-free, on Dartmoor or in the New Forest, but even here doped feed can be left out for them if the fly ever came to be regarded as a local nuisance again.. Although this is not a dung fly it is, nevertheless, a fly associated with dung, but for how much longer? The veterinary chemicals given to farm animals will have a darker, more profound effect in chapter 10.

THE MYSTERY OF THE DEEP

What route the tiny shining spider beetle *Gibbium aequinoctiale* took into the deep underground passages of the Silverwood Colliery on the outskirts of Rotherham remains unknown. They were discovered here in the 1990s. Coal miner Andrew Constantine noticed them, because his brother Barry was an entomologist, and he knew that anything eking out a living 800 m below sea level in the dark, disused tunnels was bound to be something unusual. It took several years, though, before he could be convinced to collect any specimens, because the beetles only occurred in the old roadways that the miners used as latrines, and here they were feeding on human excrement (Constantine 1994).

A deep-shaft coal mine is not your average workplace; there are no lavatories, toilets or restrooms, no plumbing, no water, no sanitation of any kind. Down at the coalface for hours on end, the miners took whatever convenience they could when the need arose, so they used the abandoned roadways as unofficial earth closets, except there was no earth down there either. The walls were bare coal, the floors crumbly mudstone. Unlike latrines on the surface, which might be leached by rain, visited by dung beetles and flies, and eventually absorbed back into the environment, down in the cold coal underworld no dung recycling took place. Instead, the stools slowly began to dry out until they reached the consistency of fruit cake. This is not a faecal incarnation represented on the Bristol stool chart. It was in this almost non-decaying organic matter in which the beetles were breeding – hundreds of 'em. Since this unusual habitat was first noted, specimens of the beetle have also been found in similar circumstances in coal pits in Staffordshire and Durham.

Several species of spider beetle are known to feed on fallen seeds, mould, pollen stores in bees' nests and stored foods in human habitations. They are adaptable feeders, and many have become cosmopolitan household pests by invading our larders. It was suggested that down the pit they may have been feeding on spilled food, but who eats their sandwiches in the lavatory? As Mr Constantine observes: 'The miners understandably take their meal breaks well away from these roadways, so there is no food debris for the beetles to feed on.' Pit ponies were used in the mines from their opening in the 19th century until about the late 1960s, so one scenario is that the spider beetles somehow arrived with the ponies' feed and straw bedding. However they got there, the Silverwood colonies of *G. aequinoctiale* are, as far as I know, the only coprophagous members of this family.

There are plenty more deep mysteries to fathom. Where has *Aphodius subterraneus* gone? This once widespread dung beetle has vanished from the UK; the last one was recorded at Scarborough in 1954. Why is the scarce northern and western British form of the dor beetle *Trypocopris vernalis* shiny blue-black, but in southern Europe a rich iridescent green? Does anything specialise in mole dung? How did a specimen of the South African *Onthophagus flavocinctus* end up dead in a puddle on a bridle path in the middle of East Sussex in 1967? If enough people become interested enough in dung and its inhabitants, who knows what new and exciting discoveries are still out there waiting to be made.

CHAPTER 8

CROSS SECTION OF A DUNG PAT – A SLICE OF COPROPHAGOUS LIFE

IT'S NOW TIME to get the trowel out, or the stout knife, and to consider a more formal dissection of the pat. When it emerges, fresh, fragrant, glistening, extruded into the world, mammalian dung is a more or less uniform homogeneous mass. But as soon as it hits the ground it starts to change. Even runny cow dung, if deposited on a hot sunny day, starts to dry out. This has important implications for many dung-feeding and dung-breeding flies which rely on its soft moistness to feed and lay eggs.

There is a useful analogy to be made in comparing a pile of dung to a pond. The surface has its own specialist fauna. Water, by virtue of the complex physics of hydrogen bonds, has a surface tension, forming a meniscus layer, a film on which water skaters can skate and whirligig beetles whirl. The pond surface fauna has its own name, the neuston. Slightly more prosaically, dung has a rind or crust, but the division between external and internal dung insects is a good one.

Flies are the major inhabitants of the outer dung surface. The yellow dung flies, *Scathophaga*, with all their posturing and mate-guarding bravado, take up vantage points across the surface. They do not have powerful or stout egg-laying equipment, so they can only penetrate the freshest skin to lay their eggs. They are joined, albeit cautiously, by the wing-waving ant flies, Sepsidae, and the brash hubbub of green- and bluebottles. To the onlooker this is just as much a mad scramble as the thousands of dung beetles descending on an elephant dropping. It is chaos, and where there is the distraction of chaos, there is always someone waiting to take advantage.

This is where the pride of Kent, *Emus hirtus,* wades in, along with the other predatory rove beetles. Wade isn't, perhaps, the best description for the movement of these insects. These carnivorous rove beetles are agile and active, and those that hunt on the dung surface are elegant and graceful beyond the down-to-earth situation in which they find themselves. They flit about at top speed, first skulking down the side of the dung, then scampering pell-mell across the top, scattering flies much like an overenthusiastic puppy might scatter pigeons on its first walk in the park. Inevitably the first few flies in its path are too quick; they have good eyesight and rapid reactions, but the hurly-burly obscures their attack, and they soon haul off a victim. It is rapidly dispatched by the sharp scimitar-like jaws. Many years ago, when I wanted to photograph the prettily mottled greenish bronze *Ontholestes murinus,* I had to stick it in a glass tube in the fridge for half an hour, to slow it down enough to get a few shots. I resisted the temptation to spread some fresh dung on the kitchen table, and made do with a bit of garden turf for a backdrop instead.

Most of the regular rove beetle predators are just plain black, or slightly metallic bronze, but their quick movements and flighty behaviour makes them obvious and eye-catching. There runs a tale, embellished I'm sure, but more than merely apocryphal, of the late Roger Dumbrell, Sussex antiques dealer, naturalist and a bit of a character, waiting at a bus stop, when his eye was caught by a large shiny *Philonthus* rove beetle landing on some dog dung in the street. It quickly disappearing under the fresh excrement, but without hesitation he bent down, flicked over the dropping, and quickly tubed the beetle, much to the amazement, and possibly disgust, of the others in the queue. I knew Roger well enough to believe that he was quite capable of this self-assured behaviour; it was he who was partly responsible for introducing me to beetles, and the fascinating delights of dunging.

Emus and *Ontholestes* are formidable beasts, and although some of the larger *Philonthus* (*P. spinipes* at 17 mm, and *P. splendens* at 14 mm) might make a meal of a blow fly, it will be some of the smaller game they tackle most often. Cow pats are rather one-dimensional, they're really just thick pancakes, but horse droppings have a labyrinthine convolutedness offering a much more complex surface. Without needing to push or burrow, tiny flies can crawl into the natural crevices. Here they might avoid the angry scathophagids, but they must constantly be on the lookout for roving attackers.

Pulling back the upper boluses of horse dung, the inner surfaces are sometimes seething with small insects. Here are moth flies (sometimes called owl midges, family Psychodidae), lesser dung flies (Sphaeroceridae), dung midges (Scatopsidae) and fungus midges (Sciaridae). These are all tiny surface dwellers, living and feeding on the outside of the dung, but depositing their eggs on or just under the surface.

Maggots are rather more streamlined than adult flies, and with an evolutionary history of wriggling in putrid decaying matter, they are quite at home in the soft mushy interior of the dung. They do not have it to themselves, though.

SWIMMING IN THE STUFF – SOFT CENTRES

My pond analogy is tenuous, I agree, but I'm going to stretch it even further here, by suggesting that it is possible to swim through dung. Called 'swimming' dung beetles, the smooth lines of *Sphaeridium scarabaeoides* allow it to push easily through the soft molten interior of fresh cow dung. Like dung flies and rove beetles, these are rapid colonisers of fresh dung and when they land on the surface they quickly fold away their membranous flight wings and quite literally dive in. Swimming is exactly what they do, and they have broadened legs, some fringed with paddles of stiff hairs which turn their nominally running and walking limbs into oars.

Plenty of their close relatives, in the water-loving family Hydrophilidae, remain water beetles, swimming through water, crawling over submerged vegetation or clawing their way through pond-side mud. Here they feed on whatever they can find in the way of rotting detritus; some of the larger species are partially or wholly predatory.

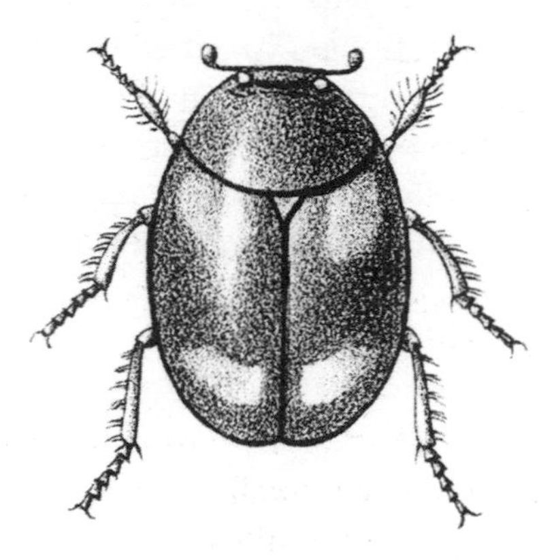

Fig. 33 Closely related to water beetles, *Sphaeridium* swims in the dung, using its legs like oars.

Sphaeridium actually swims in its own food. What a luxury. As a cow pat dries, the outer rind becomes a crust, but *Sphaeridium* is a tough insect, with powerful leg muscles and it can still push into the drying skin, long after the myriad dung flies have given up and moved on. A pat a day or so old, dry to the touch, can be distinctively pocked with small oval holes, where the beetles have continued to push their way into the still-soft interior, leaving characteristic entrance craters, as if the dung has been blasted with a small-bore shotgun.

Joining them in the viscid mire are the clown beetles (Histeridae). These smooth-lined, capsule-shaped beetles are also perfectly adapted for pushing through moist dung. Shining black, sometimes marked with vague red or orange blotches, they are sleek enough to push

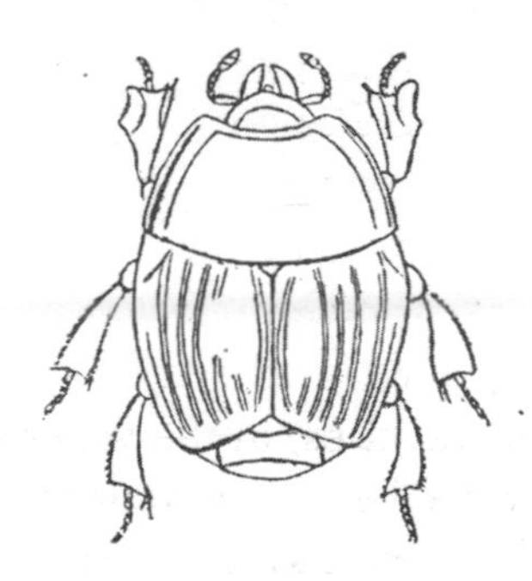

Fig. 34 Smooth lines and powerful flattened legs make histerids excellent dung burrowers.

through the soft mire using their hugely flattened and toothed legs as paddles. Both adults and larvae are predators, mostly feeding on the fly maggots which live inside the dung mass, but happy enough to take whatever comes into their jaws.

The obvious dung beetles living inside the main body of the dung are the dwellers. In most temperate zones this means the ovately cylindrical *Aphodius*, which are not so much swimming as burrowing into the dropping. Wood-boring beetles are, likewise, mainly cylindrical. With smooth heads, clean lines and broad legs, they can easily push into the dung, even when it starts to dry out. Unlike the swimmers, *Aphodius* and the other large dung beetles tend not to dive in on top of the dung, but rather push their way underneath, using the many air spaces and crevices where the dung meets the grass thatch or leaf litter. Some of the larger species such as *A. rufipes*, *A. fossor* and *A. erraticus* (the biggest UK species at 12, 10 and 8 mm, respectively) spend most of their time on the underside of the pat, whilst the smaller *A. pusillus*, *A. consputus* and *A. granarius* (all around 3–4 mm) are more able to find gaps and weaknesses in the fibrous material and push up into it.

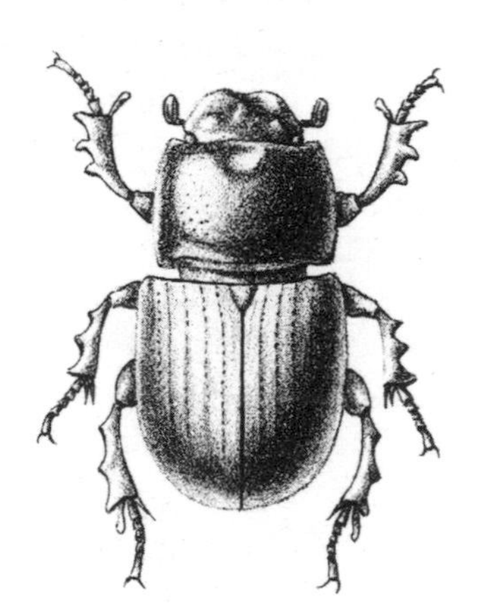

Fig. 35 *Aphodius fossor*, a sleek black monster in the dung.

It is inside the dung that the grubs of these beetles, and the maggots of the various flies, will develop. This makes the dung attractive to a whole new series of creatures. Rooks, crows, magpies and other corvids can sometimes be seen picking apart cow pats and eating the insides. They are not, of course, eating the dung itself, but are feasting on the fat larvae inside. During autumn and winter, when food is thin on the ground, two-thirds of cow pats can be pecked open by birds. Earlier on in the season, a scattered dropping is more likely a sign that entomologists have been at work. Badgers are also attracted to grub-laden cow pats; they lift off the upper portion of the pat, as if they are prising off a lid, and scoff down the tasty morsels they uncover.

THE SOIL HORIZON

The region where the dung meets the soil is not quite so crisply defined as the rind developing on top of the pat. The activities of dung beetles, dwellers living just under the pat, and tunnellers actively excavating the soil serve to blur this boundary. The shallower burrows, some little more than scrapes in the soil, may contain particles of dung which although they have been shovelled there by beetles, may still be joined, more or less remaining part of the main dung mass. Others go deeper and the fragments of dung carried off to form brood balls become satellite dung morsels. If rollers are present, the dung from a single dropping may end up being scattered across scores of square metres, at depths varying from a few centimetres to over 2 m. It very quickly gets to the point where it is difficult to know where the dung ends and the soil begins.

Despite lack of clarity, the dung–soil interface remains an important ecological zone. Here is a natural fissure along which creatures can creep. The soil delivers moisture to prevent the dung drying out too much, and the fermentation of the gently mouldering dung raises the temperature very slightly. This may only be a fraction of a degree in a single pat, but in large manure heaps the internal temperature may reach near scalding, up to 75°C. This is one of the principles on which composting works for gardeners and farmers, the heat reaching a point at which fungal spores, microbial diseases and unwanted annual seeds are killed. During my brief teenage dung beetle survey of South Heighton I was able to find beetles, even in the depths of winter, by searching in the sheltered root thatch at the centre of the gently fermenting dung. Here they remained active, or at least not completely dormant through December to February, when almost every other living thing had shut up shop to hibernate.

Just as the tunnellers are removing small portions of dung and relocating these down into the soil, so too soil inhabitants start moving out of the humus layer into what may or may no longer be part of the pat. The archetypal soil organism is the worm, and since these are ubiquitous wherever there is a rich soil humus component, they start to feed on the dung, from the underside, almost as soon as it appears. The common earthworm (also called the lobworm), *Lumbricus terrestris*, is a familiar garden species, and many others occur, but this is replaced by the distinctive reddish barred brandling worm, *Eisenia fetida*, in areas rich in organic matter such as compost bins, manure heaps and areas of the paddock where animals are regularly corralled, allowing a build-up of dung material in the soil. The lobworm remains in its mucus-lined burrow, emerging at night to grasp whatever dead plant matter it can, pulling it down into the ground and devouring it in safety. The difference between a bit of dead grass, or a bit of grass cut and chewed and passed through a herbivore digestive system, is minimal as far as the worm is concerned.

In Britain, at least, earthworms do not just play cameo roles, they are major movers. Something in the region of 30–60% of pat material is removed by them, working unseen and all but unnoticed in the hypogean dark.

Slugs and snails also move in. On holiday in France again, some years ago, I was fascinated to watch the giant edible snails, the escargots *Helix pomatia*, regularly feeding on fox droppings around the edges of the fields. The foxes had been feasting on the windfall wild plums and greengages, so their runny excrement was far more fruity-fibrous (rather than meaty-feathery) than normal for a carnivore. It was still pretty offensively aromatic, but the snails were wolfing it down. This was a warning, if one were needed, about the dangers of harvesting wild snails for the pot, unless you know what you're doing and can safely purge their digestive systems in a salad-filled vivarium for several days. I was not tempted.

Slugs, in particular, are important soil/dung invertebrates, more so than snails, which find it difficult to manoeuvre their shell encumbrances. Slugs, by virtue of being able to insinuate themselves into the narrow spaces in humus and root thatch, are very much soil critters. They are also more adventurous in their food choices; many are predatory, attacking snails and each other, or scavenging riper decay products than snails' mainly plant-based diet. Slugs are some of the main dung-feeding animals in my own back garden, where they readily attack fragrant fox and cat droppings. The high-meat diet of those animals, no doubt, produces dung more redolent of decaying flesh – to the taste of the often predatory or carrion-scavenging slugs hereabouts.

At some point, the boundary between dung and soil is completely blurred. Humus is arguably the constant churning remains of decaying organic matter and earthworm casts anyway. Charles Darwin wrote an authoritative monograph on earthworms (Darwin 1881), and came to the conclusion that in 10–20 years, they could turn over the top 15 cm of soil, eating, digesting and casting it up again. In a well-grazed (and therefore well-manured) field the gradient from dung to soil is soon imperceptible. The soil subsumes the dung. The soil is the dung.

CHAPTER 9

THE AGEING PROCESS – TIME LINE OF A DUNG PAT

ON THE AFRICAN savannah, a pile of elephant dung weighing 20 kg may have an existence measured in hours; this is recycling at its most efficient. In the arctic tundra, reindeer droppings may remain untouched for 5,000 years (Galloway *et al.* 2012), and although they can be useful in informing us what prehistoric caribou ate, for the dung fauna they are just frozen nuggets of a promise unrealised. Somewhere in the middle is the time scale appropriate for a dung pat near you.

However, even at a single fixed locality, the life span of a pat will vary. Since it is a complex community of organisms doing the dung recycling and removal, the pat moves at their pace, and this will change with the season, the temperature, the climate and the weather. Standard decay times for a cow pat in the UK vary from 35 to more than 150 days. In California they sit around for 360–1,000 days. In New Zealand 520 days are quoted. In Canada the figure is 'up to years': I suspect the dung-watchers finally ran out of patience, or their funding stopped and they had to call it a day.

Season is all important here. After the success of my May beetle-luring experiment down near Reading, I revisited the site that September and thought I'd repeat the exercise. There were a few flies, but though I waited half an hour not one beetle put in an appearance. It seems the dung beetles of Padworth were spring species. Britain does have summer and autumn dung visitors. The noon fly *Mesembrina* is widely quoted as being on the wing from late April to October, but this is a July and August insect as far as I'm concerned. I especially associate it with my childhood holidays to the Isle of Wight, Purbeck and Lyme Regis, where grazing meadows abound, and I was in a mood to chase large black flies with my home-made insect net. August and September are top months for the slightly fuzzy dwelling beetles *Aphodius obliteratus* (which quickly obliterates a pat in the hot sun) and *A. contaminatus* (which I think is unfairly named). Late autumn is my time to look for the rare *A. consputus*, which I've only ever found in October (flood refuse near Alfriston 1974, and dog dung in Friston Forest 1975). It may be 'rare' because fair-weather entomologists have retreated into their studies for the damp months of late autumn. The minotaur beetle, *Typhaeus typhoeus*, is active all winter from October onwards, and is regularly found struggling with rabbit pellets, dragging them, rather than rolling them, to a deep burrow in the sandy soil, in January and February. This marks the start of the main nesting season for this lovely insect in northern Europe (Brussaard 1983). It is one of the first beetles to fly into the moth traps of lepidopterists, when they tentatively put a bright mercury vapour light out in March.

In temperate Europe our seasons are temperature regulated, but in the tropics the permanently warm year is driven by alternating

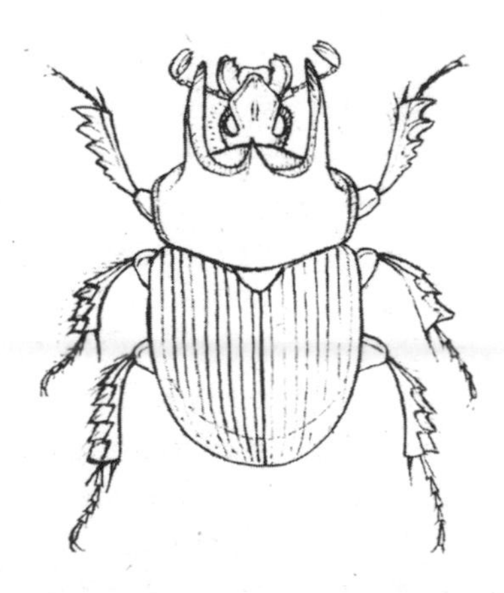

Fig. 36 The winter-active minotaur beetle, *Typhaeus typhoeus*.

bouts of wet and dry. Those fantastic reports of vast beetle numbers arriving at elephant pats are all for the wet season. This is when the beetles waiting in their subterranean pupal chambers are able to break out, pushing up through the moist soil, no longer dry and concrete hard, to find new pats. As the dry season arrives, the number of beetles attracted to dung steadily declines, until the amount they remove becomes negligible. Termites continue to remove a trickle of dung material (Freymann *et al.* 2008), but even they peter out. So rather than being gone in a matter of hours, a pile of elephant dung can easily remain for 4 months, and one monitored heap persisted for 850 days.

Exact times are somewhat irrelevant, then, but there is a slightly predictable process from newly dropped faeces, to the point where you'd never know there was anything there in the first place.

Assuming the dung boluses are not removed in ultra-quick time by hordes of eager tunnellers and rollers, a dung pat's maturation might go something like this:

fresh	hot on the ground, steaming, ready and waiting
mature	the first flush of incomers are breeding
mined	maximum occupancy of various life stages, overlapping generations
mouldy	initial attractiveness is wearing off, the pat is drying
mouldering	no longer a cohesive unit, weathering, fraying at the edges
crumbling	falling apart, pecked apart, disturbed by plant regrowth
ruins	just bits of dung, fragments, leftovers
dregs	powdery remains, a dusting of particles
echoes	no dung left, but there is evidence it was once there
ghosts	the occasional reminder
gone.	

This is my personal take, anyway, based on exhaustive researches at the dung face over the last 40 years. So this chapter now follows a nominal British or European cow pat or horse dropping from delivery to removal.

NEWLY MINTED, GOING ON MATURE

Most of the book so far has been about this stage: the first appearance of the dung and the first arrival of the creatures it attracts. It might last hours, it might last days. The waves of flies and early-arrival dung beetles get stuck in as quickly as they can. It's a time of bustling activity. This is the opportunity to sit and watch fresh dung, as the flies skitter around on the surface and the beetles plump down in the grass, then push their way underneath.

Ironically, there may be an initial negative impact on the environment immediately around the dung, especially if urine is also released. This can scorch the ground. High nitrogen levels, from the urea, uric acid or ammonium compounds, are toxic to plants, and the surrounding vegetation can be seared. I was beginning to worry about the patches of browning grass on my uneven lawn, fearing they may be some alien fungal disease, but I later spotted one of the cats squatting over it. Little monster. It is the mix of uric acid in bird droppings that makes them relatively unpalatable to dung-feeding insects, and a reason to dismiss dinosaur dung beetle claims. The cats, though, defecate elsewhere, in the flower beds, so visiting dung beetles, often the attractive *Onthophagus coenobita*, are not too put off. Cows and horses also tend to release their different bodily wastes in separate motions, so dung released in the open space of the grazing meadow is unlikely to be polluted with urine.

This fresh stage may last a few days or a couple of weeks in sunny southern England and is, to my mind, the most productive and interesting phase. The dung maintains its softness, so is easily manipulated when turning it over or cutting it open. This is the stage where the dung fauna is made up mostly of adult beetles that have arrived to feed and breed, and every dropping might hold some new treasure hidden beneath it.

There is a tale, based in truth I'm sure, but clouded by the fact that several slightly different versions have been recounted to different listeners. My old friend Roger Dumbrell told of a beetling trip where he and A.N. Other (name withheld by request) went to Camber Sands, near Rye, in East Sussex, back in the early 1970s before I really knew him. Walking onto the dunes from the car park they passed a large dropping on the sand and paused to consider whether or not to examine it. It was obviously fresh and the used tissue close by gave no doubt that this was a human discharge. One of them walked on, unwilling to engage that particular taboo, but the other gingerly turned the dung over with a stick. The beetles underneath were the very rare *Onthophagus nuchicornis*, and as the cry went up the previously unwilling coleopterist shelved disgust and came back to join in the exploration. Who walked on, and who first dived in, is now lost in the debate fogs of hazy memory, but the specimens remain in the collection, dated 4 May 1972, recorded as being under dog dung. Mind you, this is the same Roger Dumbrell who refused to write 'under an old shoe' in his catalogue, when he found a rare ground beetle under flood refuse in a saltmarsh; instead he wrote 'under rejectamenta'. There was a time when an entomologist would write 'in stercore humano' in a published article, hoping the use of a classical language would somehow excuse or disguise his questionable behaviour. Incidentally, the name *Onthophagus* derives from the Greek ονθος *onthos* 'slime', and φαγειν *phagein* 'eating', particularly apt for a group of beetles often found under the riper droppings of omnivores and carnivores.

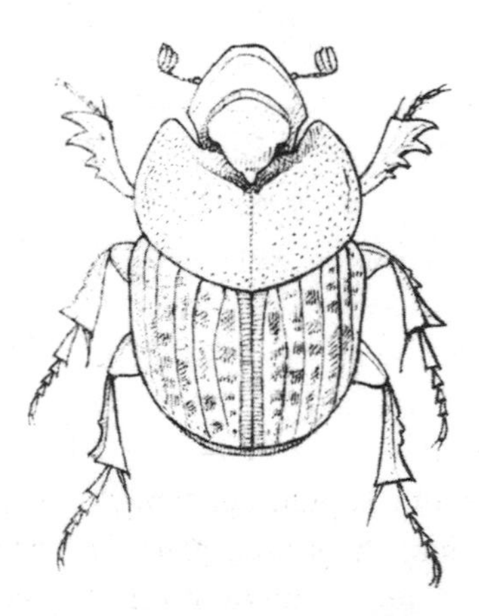

Fig. 37 *Onthophagus*, a name meaning 'slime-eater' occurs in some of the riper, more fragrant dungs, including those of cat, dog and human.

Whatever the origin of the dung, the first few days are the time for colonisation. First come the dung-feeders, with numbers peaking around the second day. After about 3 days predator numbers increase, peaking about another week later. In a famous study of beetle succession in Finnish cow pats (Koskela and Hanski 1977), 10-day tallies showed there were more carnivores in the dung (average 191 individuals, of 16 species) than coprovores (126 individuals, 11 species); but the biomass (total weight of all the beetles together) was greater in the dung-feeders (443 mg) than the predators (129 mg). The later arrival of the predators makes evolutionary sense; if you are a carnivore, better to wait until there is a thriving bustle of colonists on the dung before launching in on the attack. The study did not measure numbers or biomass of fly maggots, but since most predatory dung beetles are maggot-eaters, it also makes sense to wait until at least the first generation of fly larvae are up and running.

THE WELL-DEVELOPED COMMUNITY

After the initial influx of new colonists, by about the second week of lying in the field, the pat no longer attracts hordes of new visitors. It arrives at a sedentary phase, where the developing larvae and maggots feed quietly within. Adult beetles are still present, but there is also movement off to new pats. The outer layer starts to dry, but the interior remains moist. This is the stage at which dissection skills really count.

I now look back on my early dunging life and sometimes regret the carefree exploration of so many pats teeming with life. I was careful to look for new and unusual species, but found the great numbers of common fry a mere distraction. This was a time of community dunging, when I might be in the company of Messers Dumbrell and Hodge, my father, and possibly other budding entomologists. When three or more of us surrounded a heavy horse dropping on a cliff-top path near Brighton, on 14 September 1974, I was pleased to note two different colour forms of the scarce *Aphodius scybalarius* (now called *A. foetidus*), and the equally uncommon *A. foetens*. But I did not count the exact numbers of what must have been of the order of 200 specimens of several commoner species, or make any assessment of the multitudinous larvae squirming in the fibrous mass. Thankfully other more mindful entomologists have gone into the numbers. The Finnish cow pats of Koskela and Hanski (1977) yielded a total of 62,498 beetles, of 179 species, gathered from 312 carefully arranged 1.5 kg dung masses. Coincidentally that's about 200 specimens per pat, so I'm relieved that my memory appears to serve me well.

This is the time of greatest invertebrate diversity in the well-mined dropping. Although less obvious in temperate Europe, subtropical pats soon start to show a skewed sex ratio of dung beetles; males dominate the upper pat as females become increasingly subterranean, buried with their dung brood balls.

The single entity extruded at the beginning of the process has developed subtle moisture and temperature gradients, and each dung insect chooses its own very particular niche. There are huge size differences, and even in a British pat, with our impoverished island fauna, the heaviest dung beetle (one of the dors) can be 5,200 times as heavy as the lightest feather-winged beetle (see below). Tiny insects dominate, and despite 200 *Aphodius* dung beetles seeming a lot to me, this is nothing compared to a recorded 20,000 owl midge larvae in a single dropping. This diversity does not last long, though.

The early arrival of fast-breeding flies and get-in-there-quick beetles reaches peak activity in the first fortnight, but tapers off quickly thereafter. By about 8 weeks, after being riddled with adult and larval activity, and following sequential wetting and drying in Europe's notoriously changeable weather, the dung is starting to look more fibrous; the pat is no longer a cohesive whole, but it starting to fray around the edges. Cattle dung often starts to show a bloom of the small orange ascomycete fungus *Cheilymenia fimicola*. This is the beginning of the end.

THIS PLACE IS FALLING TO PIECES

By this point, the familiar dung odours have evaporated, the bacteria that made up so much of the dung have been eaten, or gone into quiescent spore mode, and the dung really does look just like mushed up hay or processed plant material. There may still be a few dung beetles hanging around, but this is at the tail end of their interest. This is now a time for fungus.

Most of the fungi feeding in dung are small (microscopic really) and poorly understood. There are mind-boggling numbers of them. One millilitre of sheep dung can contain a million fungal mycelia (each mycelium being the fungus equivalent of a root system), and a gram of cow dung can have two million yeast cells, yeast being single-celled fungi. Despite a common misconception, fungi are not plants, they are a separate group of organisms. The best way to consider fungi is think of them not as producers (like photosynthesizing plants) but, like animals, as consumers. However, unlike animals, which do their digesting on the gastrointestinal inside, fungi do their digestion on the outside, absorbing nutrients with their tendril, root-like hyphae.

Many of the fungal spores are already in the dung when it is delivered, having been resting on the plants eaten by the animal in the first place. They remain inert, passing unharmed through the herbivore's digestive tract until they are voided and can get to work on the dropped dung. The fungi are encouraged by the burrowing of the beetles aerating the dung. Occasionally they will throw up a fruiting body to produce spores. Most of these fruiting bodies are also microscopic – moulds and mildews – but the odd toadstool sometimes appears. For those who are interested, there is a useful guide to British coprophilous fungi (Richardson and Watling 1997); the authors suggest incubating sample material on a table in a warm room, between sheets of absorbent paper, or perhaps maturing in a glass casserole dish or plastic sandwich box. In his book on British cup fungi, Dennis (1960) exhorts: 'A rich harvest may well await the man who cares to devote his leisure hours or his declining years to a study of stale dog dung.' And you thought entomologists were disgusting?

Dung lying in the fields also sprouts up fruiting bodies. These can be highly distinctive, such as the delicately sheened egghead mottlegill *Panaeolus semiovatus*, or the more robust dung roundhead *Protostropharia semiglobata*. My favourites are the ink-caps, several *Coprinus* species. These large, distinctive mushrooms, with tall, elegant, almost cylindrical domed hoods, get their English name from the black dripping inky deliquescent mess which they become after producing spores, but the scientific name is obviously from the Greek word for dung. We used to find these often, when out on family rambles, and my father would collect them for a tasty snack on toast later that evening. He could eat them with impunity, him being a teetotaller, and all. Be warned, though, if you ever have these succulent fungi with a glass of wine, you'll feel pretty bad afterwards. They contain a compound called cyclopropylglutamine (sometimes also called coprine) which blocks the enzyme acetaldehyde dehydrogenase, part of the natural metabolism of alcohol in the human body. This leads to a build-up of acetaldehyde, one of the chemicals that makes 'off' wine smell unpleasant, and which is often blamed for producing hangovers. Symptoms of the resultant aldehyde poisoning include facial flushing, nausea, vomiting, palpitations and general malaise – exactly like a hangover.

Meanwhile, back in the dung, the presence of countless fungal bodies and spilled spores now attracts another host of insects, this time fungivores. To be honest, for many of the smaller insects, it is not actually very clear whether they and their larvae are feeding on the dung, microbes such as bacteria, or on the fungi. These are mostly very small fry indeed, microscopic feather-winged beetles (family Ptiliidae) half a millimetre long, ridged rove beetles (*Micropeplus* species), tiny globular mould-feeders such as *Atomaria* and *Ootypus* (family Cryptophagidae), and the tiny larvae of very many groups of fungus gnats (families Sciaridae, Mycetophilidae, etc.).

Outwardly, the dung is on its last legs, crumbling to pieces, becoming powder even. Plants are growing up through it, and it is very quickly being reabsorbed back into the landscape. It is soon very little different to the other general leaf litter, so it's no surprise that general leaf-litter organisms start to invade. Here are multilegged millipedes and centipedes. The gnarled flat-backed millipedes,

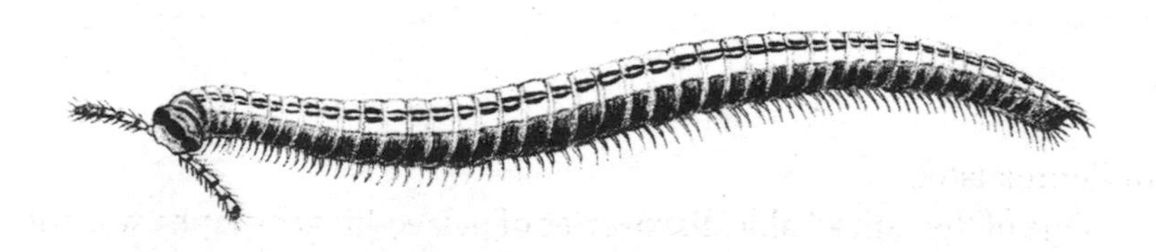

Fig. 38 Cylindrical snake millipede, shaped for borrowing in the rapidly blurring dung–soil horizon.

Polydesmus, are probably genuinely feeding on the dung, they also occur under rotten logs, chewing at the fungoid decay. The shining black snake millipedes, things like *Tachypodoiulus*, are general leaf-litter scavengers, but they regularly appear as the soil–dung boundary becomes less defined. The centipedes are predators, hunting after whatever small soil-dwellers they can get; these will be various soft-bodied soil-dwelling invertebrates such as springtails, and bristletails, but also any lingering fly larvae.

Before it finally gives up the ghost, the dung, now in pieces, can be home to just about any soil surface inhabitant. Woodlice may feed, but they are also there for shelter. Root-feeding insects such as click beetle larvae (wireworms) and crane fly larvae (leatherjackets) make little distinction between tough cellulose grass roots and the final fibrous cellulose chaff from the dung. Ground beetles (Carabidae), smooth, sleek shiny predatory species, push through the root thatch, and have specially muscular back legs for wedge-pushing into tight spaces; they end up hiding and hunting under dung, just as they would under a rock or a log.

Ants feature little in most books on dung ecology, but they are a regular feature of older horse droppings. Soil-nesting species may be attracted by moisture if the weather has been warm and dry for some time; predatory species may well be hunting after small prey items.

VERY LITTLE LEFT NOW

There comes a point at which the dung is no more than a memory. The grass grows a little greener, but there is no physical evidence that a pat ever lay on top of it. Forensic sifting of the soil might reveal a few fragments of insect remains, bits of beetle left over from predatory attacks or subterranean breeders unable, for one reason or another, to escape the final binds of their interment. These remains last a long time, and they continue to remind us what was in the dung, even millennia later.

One of the remarkable discoveries of palaeo-biogeography was the remains of 150 subfossil specimens, and bits of specimens, of a dung beetle from peaty deposits in a gravel pit at Dorchester on Thames near Oxford in 1973. They proved to be *Aphodius holdereri*, a dweller dung beetle now restricted to the Tibetan plateau, 3,000–5,000 m on the north side of the Himalayas (Coope 1973). This is about as high as dung beetles get anywhere on the globe. The Oxford deposits, together with a further 14 sites in England where this species later turned up, are dated to the middle of the last glaciation, 25,000–40,000 years ago. With a change in the climate, and the retreat of the ice, this particular cold-adapted dung beetle was not able to survive in Britain, where it was once obviously widespread (feeding in caribou droppings?), or anywhere else in the world, except the extreme centre of Asia, where it clings on in yak dung, or whatever it can find.

That beetles, at least, remain buried in the ground long after the dung has vanished is obvious to anyone who has ever seen the rains arrive in north Africa. It was here, 4,000–5,000 years ago that the marvel of the sacred scarab was first observed, or first revered. Though they have spent many months chewing away inside their dung balls, the scarab grubs have transformed into pupae and finished their development, but remain quiescent, waiting. They are, quite literally, entombed in a rock-hard nodule, and the dry ground above them is compacted and solid, sun-beaten and parched. They will only emerge when rainwater has softened the impacted clay, pushing up through the now blank and featureless soil, erupting through the surface of the ground, as if by spontaneous generation. But there was always the danger they might never escape.

At Poona, western India, in June 1826 Lieutenant-Colonel W.H. Sykes was having some of his palanquin-bearer servants dig and break up hard ground with pick-axes to spread on his garden paths, when they encountered 'some depth below the surface, four hard perfect balls. At first they considered them stone cannon-shot.' The Colonel, however, recognised that a broken ball contained the crushed remains of an insect pupa so he kept two of them in a tin box in his study, to see what would emerge. He soon forgot about them, but on 19 July 1827 he heard an eerie low scratching sound coming from the abandoned tin, now stuffed on top of a bookcase. He realised that some beast was trying to extricate itself from the hard clay ball, but by one o'clock in the morning it had not succeeded so he went to bed. It continued its ineffectual scratching all the next day too. Finally, before heading off to his bedroom, he doused the ball in water and at sunrise on the 21st he found a fine specimen of the glossy black *Heliocopris midas*. The other emerged on 4th October. They had been in his possession for 13 and 16 months, respectively, and he estimated that they had been underground for three years before they were dug up (Sykes 1835).

These ghosts of dung life past can still be located well after the dung, and the dung-droppers, have gone. Large fossilised dung balls, 20–50 million years old, are known from South America, even though the mammalian megafauna that provided the raw materials, are long extinct – huge armadillo-like creatures the size of small cars, giant sloths bigger than polar bears and bizarre elephantine hoofed creatures. No fossil beetles are known here, but their empty rolled and buried dung balls are distinctive enough for several *Coprinisphaera* species to be named in the scientific literature. Some of the dung balls are whole, suggesting that their inhabitants never emerged. Others show the belated chewings of other creatures, cuckoo parasites, enough for these also to be named; several *Tombownichnus* species were most likely kleptoparasitic beetles, and *Lazaichnus fistulosus* looks to have been a worm of some sort (Sánchez and Genise 2009).

Back in the modern world, other secret contents in the dung balls, whether rolled away or buried beneath, are the seeds which the original herbivore (or frugivore) swallowed, and which were passed in the primary dropping. Unbeknownst to the busy beetles, they have nearly finished assisting the plants that started the whole nutrition chain in the first place. Just like the Spanish dumbledor, *Thorectes*, which helpfully buries acorns well away from the overshadowing parent tree, so too the much smaller dung beetles have taken around 25% of the seeds in the dung and neatly buried them carefully a few centimetres deep in the rich fertile soil (Shepherd and Chapman 1998; Beaune *et al.* 2012). Here the seeds escaped the attentions of sparrows, harvester ants and spiny pocket mice and are perfectly placed to germinate in months or years to come, to found another stand of tasty plants for future grazers, and future dung providers. It's a circle-of-life kind of thing.

Back at our nominal European cow pat, a year is as good a suggestion as any to declare that all outward signs of the original dung are gone. But the buried pupae of tunnellers and rollers are probably somewhere nearby. When we uncovered dor beetle pupae in the shallow soil of the Iron Age archaeological dig above Newhaven it took a few moments to realise what they were. Although many of the meadows round about were grazed, there were no animals where we were, nor were there any outward signs of their faecal signatures. Had we not disturbed them, the dumbledors would have clawed their way up through the soil a few weeks or months later, and made the short journey to the next field, where cows still chewed the cud, and where fresh pats were awaiting attention.

In moist, rain-drenched Britain it is easy to forget how rock-hard the soil can get when it really dries out, but the ancient Egyptians were right to be in awe of shiny black scarabs disgorging themselves out of the soil when the crop-giving rains returned in October. Whether they associated the beetles with a circle-of-life seasonality of crops and harvest, grass growth and grazing meadows, or the mystical daily cycle of the sun rolling through the heavens, it was something they wondered at, and applauded. What, then, would they have made of the modern-day decline of dung beetles? Could they, horror of horrors, contemplate a world where the dung beetle was no longer pivotal in the cycling and recycling of the world's ordure?

CHAPTER 10

DUNG PROBLEMS – THE END OF WORLD ORDURE AS WE KNOW IT

WHEN CHARLES DARWIN toured Van Diemen's Land (Tasmania) in early 1836, he showed his true spirit as a coleopterist by finding in cattle droppings two species of *Onthophagus*, two of *Aphodius* and another unnamed dung beetle genus (Darwin 1839). He was bemused to note that cattle had only been introduced, with European settlers, some 33 years earlier. He was particularly struck by the fact that native mammals, like kangaroos, had very different dung to cows, and that the transition from marsupial dropping to ungulate pat was highly remarkable. He was right.

The first cows (along with sheep and horses) to arrive in Australia were delivered in January 1788 when the First Fleet arrived to found the British penal colony at Botany Bay, Sydney. Two hundred years later, by the end of the 20th century, there were coming up for 25 million cows, 70 million sheep and 200,000 horses. This is a considerably higher biomass than the estimated 50 million kangaroos, but all would have been well if they had shared a similar digestive metabolism, and produced similar dung. However, as Darwin had noted, the native Australian fauna produced droppings very different indeed from the copious discharge of the colonists' stock animals.

It all started about 100 million years ago (MYA), when the tectonic plate carrying Australia started drifting away from Gondwana. Dinosaurs were still the dominant land animals at the time, and presumably the major dung producers. We're not quite sure exactly what happened when they vanished 65 MYA, but marsupials have been the dominant dung deliverers in Australia for the last 25 million years at least. Until the Europeans arrived. Kangaroos (along with wombats, koalas and other herbivore marsupialia) are well adapted to the arid, often drought-locked continent, and their droppings are dry, tough, hard nuggets. Having evolved for so long to this dung format, the sudden arrival of semi-liquid cow pats was about as alien to the native dung beetles as you could get. Darwin was right to be amazed at this transition, a few of the native species found a useful new ecological niche; but only a few.

There are currently thought to be around 600 species of native Australian dung beetle,[1] with most found in woodlands and forests. The savannah/ grassland habitats are poorly utilized by native herbivores, so few dung beetles occur in this habitat (Doube *et al.* 1991). But these are exactly the areas 'improved' for grazing farm animals. The dung fell, and it was all but ignored by the majority of native dung beetles. It was not ignored by the flies.

A FLY IN THE BUSH IS A PAIN IN THE EYE

As an entomologist I am very tolerant of insects, and refuse to dismiss them as pests... unless they reach pest proportions. Anyone who lived in the Australian outback towards the end of the 19th century could certainly tell you that bush flies had become pests; there were vast black clouds of them, like smoke, and it was not possible to speak without a fly flying into your mouth. They congregated around your eyes, trying to drink the tears, they tried to climb up your nose; they got into your hair, under your shirt, in your ears. They were

[1] Scarabaeidae 437 species, Aphodiidae 127 species, Hybosoridae 40 species, Geotrupidae 1 coprophage species (Ridsdill-Smith and Edwards 2011). This does not include the very many other beetle species that occur in dung, but as has been reiterated enough times already, it is these groups in the superfamily Scarabaeoidea that do the lion's share of the dung removal and burial.

everywhere. They were horrid. These were pest proportions on a biblical scale.

The bush fly, *Musca vetustissima* (meaning the 'inveterate' fly), is nearly the Australian equivalent of the Eurasian house fly, *Musca domestica*. It's of a similar size and shape, but it does not tend to come indoors so much; it's everywhere else instead. Like the house fly, the bush fly breeds in decaying organic matter, and it just loved the new dung. As a rough estimate, farm animals now drop approximately 2 million tonnes of dung each day onto the grasslands of Australia. With each 2 kg cow pat producing a thousand or more bush flies, populations rose exponentially. The stable fly, *Stomoxys calcitrans*, and the buffalo fly, *Haematobia irritans*, were quick to get in on the act too. They bite, and although their blood-sucking habits are not associated with the spread of human diseases, they can raise a painful welt, and they drive farm animals to distraction. The bush fly remains an iconic, if undesirable, Australian animal; it was to combat its attentions that bush hats decorated with corks dangling on string became a staple fashion accessory in the outback, or on TV sitcoms, and its persistent visitations to the eyes and nose gave rise to the Aussie salute, a quick flick of the hand across the scowling face.

Cows, in particular, were the problem. Their wet dung was incomprehensible to most native Australian dung beetles, used to the hard, golfball-sized droppings from kangaroos and the like. And there was just so much of it. Despite the attentions of a few species, like those found by Charles Darwin, native dung beetles just could not cope with the removal; they, and the grasslands now being farmed, were overwhelmed. Instead of being digested, buried, removed, recycled, the pats just sat there, drying to hard pancake crusts in the unrelenting Australian sun. Even during the winter, no noticeable degradation took place and a pat laid one year looked very much the same 12 months later. Two oft-quoted statistics are that the dung of five cows can smother an acre of meadowland in a year, or the dung of ten cows can smother a hectare. These don't quite marry up mathematically, it ought to be 12.3 cows to cover a hectare in pats, or 4.1 cows for an acre, but that's in the nature of easily quotable tabloid statistics. But you get the idea. Since the arrival of farm animals with the First Fleet, Australia had been gradually disappearing under excrement. More shocking statistics are offered up: meadowland was being rendered useless at a rate of 200,000 hectares (2,000 km^2) per year. Parasitic worms, whose cysts and eggs are spread in the dung, were not being controlled by dung removal, so reinfestation rates soared. By the 1960s, after 200 years, things were getting desperate. It took the arrival of an entomologist with an interest in scatology to get things on the right track.

BEETLES TO THE RESCUE

When Hungarian zoologist Dr György ('George') Bornemissza arrived in Australia in late 1950 he immediately noticed that the Australian meadows were disfigured with undecaying cow pats lying about all over the place, completely in contrast to the well-maintained cattle fields he'd grown up with back in Europe. He quickly deduced that what was missing were the hordes of dung beetles that should normally be digging in and burying the stuff. Obviously. So was born the Australian Dung Beetle Project. Its aim was simple: to introduce into Australia some non-native dung beetles, species which had co-evolved with the non-native cattle, and which would be quite at home in the cow pats. This should have the quadruple benefit effect of physically removing cow dung so it did not clog the fields; increasing the nutrient cycle to improve grass quality for better milk and beef production; containing or reducing intestinal worm reinfestation in the cows; and decreasing the breeding grounds of the noisome bush flies and its biting confederates (Ridsdill-Smith and Edwards 2011).

The trouble is, if you introduce an alien organism into a new continent, you never really know whether it will play havoc with the endemic plants and animals that have already evolved through millions of years without it. Australia had already suffered huge environmental damage from the cane toad, *Rhinella marina*, introduced from South America in 1935 to try and keep down equally exotic insect pests in the sugarcane fields of Queensland. It ate the pests, but also native animals, many of which it has now driven to the point of near extinction. With poison glands behind its head and with no natural predators, the cane toad has itself spread widely and in huge numbers, becoming an invasive pest in need of control. Foxes, cats and rabbits had similarly decimated Australian wildlife.

Bornemissza's aims were true, but great care, and a deal of good public relations, would be needed to try and avoid similar environmental upsets with foreign dung beetles.

It would take some years of careful plotting to select and trial dung beetles for eventual release. Sponsored by the Australian government and funded partly by the beef industry, the project officially began in 1964. As a template, dung beetles were first collected from Hawaii. These were not originally endemic to the Pacific islands, but had themselves been introduced as part of a similar project to try and control buffalo flies and horn flies – insects that had achieved pest status in Hawaii for the same reasons as the bush fly in Australia.

The first beetles to be released (in 1967), namely two clown beetles *Hister* (*Pachylister*) *chinensis* and *H. nomas*, were not dung-feeders, but fly larva predators. This was almost as an aside, even as they were being let out it was realised that dung removal by true dung beetles would be the only way to try and even out the ecological imbalance caused by Australia's megatons of cattle dung.

To maintain biosecurity, i.e. to prevent unknown diseases from arriving, no adult beetles were imported; instead surface-sterilised eggs (3% formalin) of 52 dung beetle species were shipped to Australia; 43 species survived captive rearing through several generations in laboratory conditions. Captive breeding is enough to distort population genetics, such that when final release into the wild came round, alterations to the vigour of the species may already have been bred in. To give the chosen colonists a good chance, some were rather pampered – released into crude field cages to prevent predation by birds, and supplied with fresh dung for a week. In the end, though, it was just a case of opening the gates and letting them out into the badlands, hoping they would do alright.

Between 1968 and 1984, 1.73 million dung beetles of the 43 different species were released into thousands of Australian meadowlands. The first of these (1968) was the African *Onthophagus gazella*, which has now become the dominant cow dung species in the subtropical areas of northern and eastern Australia. Not all of the introductions were successful. Of the 43 release species, 23 have become established. Australia is a far-flung continent, unique in much of its flora and fauna, which has evolved in the last 100 million years, adapted perfectly to its peculiar climate and geography. In sourcing the original target species Bornemissza and his colleagues had to try and work out which world species might be able to slot into the brave new Antipodean dung ecology they were planning. Tropical South African species were tried in tropical northern and eastern Australia, whereas slightly more temperate-adapted species from the Cape Province and Southern Europe were tested in southern and western Australia.

Success was fickle. On the whole, smaller species (<13 mm) did better than larger species (>15 mm), and part of this is no doubt down to the fact that the larger beetles are usually the ones to spend more time caring for low numbers of offspring, so their fecundity remains low, and their increase slow. It was no surprise, really, that *Onthophagus gazella* should thrive – nearly half a million individuals were released at 421 sites. Other large-scale releases of *Euoniticellus intermedius* (248,637 specimens, 443 sites), *Onitis alexis* (186,441 specimens, 251 sites) and *Onthophagus binodis* (173,018 specimens, 178 sites) were likewise successful.

At the other end of the number scales, only 53 specimens of the elegantly long-legged *Sisyphus mirabilis* were set free, at one site. It was probably doomed from the off. A neat table produced on the 40th anniversary of the programme's start (Edwards 2007) would seem to indicate that a minimum release of 8,000 beetles, at six or more sites, with at least one site getting 500 specimens, is the distinction between secure establishment and not. Even so, there are surprises. The large *Copris diversus* vanished without trace, though 17,775 were sent off in nine localities; however, 294 of its congener *Copris hispanus* did manage to secure one of its two release sites. These are slow-developing species, with limited numbers of offspring. The African *Copris elphenor* was released throughout Queensland, but remains confined to one small area around Jambin 35 years after its first liberation.

Success for some species was an understatement. Freed from natural predators and parasites a select few aliens did exceptionally well. In recovery sampling measurements, the biomass (total living weight) of African beetles from dung at Rockhampton, Queensland, was 10–100 times that back in Hluhluwe, South Africa, from where they originally came. There has been some field collection and re-distribution of the successful colonists. Between 1989 and 1995

3.5 million beetles were caught using baited traps and shipped off to other sites. By this time funding from the beef producers had run its course, but the dairy industry stepped in to keep things going. Since 1993 private companies have collected, bred and re-released over 6 million beetles, shipping them by the thousand, and charging farmers to supply the meadow clean-up brigade.

Once established at a locality, many of the commoner beetles started to spread of their own accord. *Onthophagus gazella* was off like a gazelle, extending its range by 50 km/year, and making one leap of 800 km. It was able to colonise offshore islands, flying over 30 km of open sea. This is on its way to becoming one of the most cosmopolitan insects in the world, and has also been deliberately released for dung clearance work in Hawaii (1958), North America (1972), South America (1990) and more recently to Easter Island, New Caledonia, Vanuatu and New Zealand. The southern European *Onthophagus taurus* spread at 300 km/year, but has now more or less reached its climatic limits in western and southern Australia and Tasmania. A specimen was found in the belly of an Australian herring, *Arripis georgianus*, caught in the Great Australian Bight (Berry 1993). Most species, though, increased their ranges at a more sedate 1–2 km/year.

A species survival success rate of 54% (23/43) is still pretty impressive. Edwards (2007) is proud to claim that at least one species of introduced dung beetle is now firmly established in every grazing meadow in the country, with a maximum (so far) of 13 exotic species at Toowoomba, Queensland, where several tropical and temperate species overlap. But have any of the anticipated benefits been achieved?

Despite nearly 50 years of work, this is still early days in a terraforming project to change the ecology of an entire continent. Things are looking up, though. In field experiments with cow dung, an average of only 2 bush flies emerged from pats populated by the South African *Onthophagus binodis*, compared to 128 when the beetles were excluded. That's a tabloid statistic of 98.4% reduction. Seemingly a mere 3 g of dung beetles per pat was enough to reduce the flies' nuisance. This is a real advance on the paltry 33.3% reduction achieved when only the native *Onthophagus ferox* were present

Removal of cow pats has certainly speeded up. Ball rollers such as *Sisyphus spinipes*, working thousands to a pat, destroy it in a day.

And even though this species does not bury its balls (it just rolls them away and anchors them to bits of vegetation), dispersal of the single dropping into lots of tiny bits, each then being devoured by a beetle grub, still achieved the desired result.

Nutrient cycling has also improved. Before the exotic beetles appeared, only about 2 kg of the potential 13.5 kg of nitrogen excreted in the dung of one cow during a summer was returned to the soil. After beetle inoculation, 10 kg is now divested to the hungry grass.

The effects of dung beetles on the cows' intestinal parasites, spread through cysts or eggs in the dung, then reingested as they spread onto next year's grass, is still in the balance. In at least one field experiment nematode worm larvae were reduced when the heroic *Onthophagus gazella* was present, but rainfall had a significant effect too. There is still work to be done here.

On the plus side too, fears that these introduced exotics might interfere with fragile native ecosystems seem to have been calmed. There was always the possibility that incoming dung beetles might outcompete the endemic species. We cannot quite be sure yet, but it seems that the original Australian dung beetles, mostly living on marsupial pellets in the forested areas, continue to mosey along in what is, in effect, a parallel rather than an overlapping ecosystem.

This was always going to be a long-term project. If it took 200 years of European-style farming to create the problem, it might yet take a few more decades to clear things up. It's just possible that a similar tectonic ecological shift occurred on Madagascar 1,000 years ago. This was when cattle first arrived to an island that for the previous many millions of years was dominated by lemurs and their relatives. Cow dung is not so much of a nuisance as in Australia, but there is still evidence that domestic animal dung (7 million cows, 1.5 million goats and sheep) is used by only a small number of the Malagasy endemics which are still rather adapted to the small droppings of the native forest-dwelling primates.

The last 50 years have been spent concentrating on Australia's excess cattle dung, but sheep droppings, being harder and drier (but not up to marsupial standards), require different species, and this still needs to be addressed. Throughout the rest of the world individual grasslands and savannahs can have a dung beetle diversity of over 100 different species. Additional new species were released in 2014, but Australia's meadows are still some way off this figure, so there is certainly room for more introductions in the future. Most Australian dung-beetlers already have their wish-lists ready.[2]

AN IMPENDING ECOLOGICAL DISASTER OF OUR OWN MAKING

Dung beetles and dung flies (as long as they don't reach pest proportions) are the unsung heroes of the environment. Without them, as was shown in Australia, we would be wallowing in our own or our farm animals' ordure. Unfortunately the world has changed since the beetles were worshipped and celebrated by the ancient Egyptians, and now we take them for granted. Indeed, we care so little of them that we are happy to poison them. This is the unfortunate side-effect of our demand for cheap meat brought to our supermarket shelves by intensive farming.

One of the problems with free-range cows is that they tend to eat whatever is growing out of the ground. Mostly this is fresh green grass, but occasionally they swallow an egg or a cyst of some internal parasite such as nematode worms or flukes. These then live inside the cow's warm, comfortable, protective and highly nutritious body. Along with horse bot fly larvae living inside horse intestines (see page 142), warble fly maggots puncturing and burrowing under the skin of cows and leaving suppurating wounds, larvae of the sheep bot eating the poor animal's nasal passages, and a whole host of other pests, farmers struggle to keep their animals healthy and profitable. It is no surprise that they give them medicines to counter these noisome attacks.

Ivermectin is a broad-spectrum antiparasite formulation, an 'endectocide', an internal pesticide drug. It has a complex multicyclic

[2] We must, however, not imagine for a moment that there can be quick fix to nuisance dung of all types. In 2008 entomologist Maria Fremlin was intrigued to receive a request from a Toronto university student who had the idea of trying to market a biodegradable doggy poop-scoop bag with dung beetle eggs already inside, to feed on the dung cleared up by owners from Canadian pavements. Well intentioned as it may have been, once she pointed out the intricacies of dung beetle ecology nothing more was heard of this ill-conceived notion.

molecular structure (for the chemically minded it's a macrocyclic lactone 22,23-dihydroavermectin-B_{1a}), and was originally derived from a bacterium *Streptomyces avermitilis* by Satoshi Omura of Kitasato University in Tokyo and William Campbell of the Merck Institute for Therapeutic Research in 1981. It is the number one treatment for river blindness in humans, which is caused by a microscopic blood worm. It is very effective. It is also given to all manner of farm animals to counter the parasites that would otherwise blight them.

Ivermectin is delivered to the stock animals in various ways: injection into the skin, as an oral gel bolus, mixed into feed, or as a drench over the skin. However administered, it gets absorbed into the body tissues of the animal. It also gets passed out with the dung, barely changed. After a pour-on formulation was used on young beef calves (0.5 mg/kg body weight) the first day's dung contained the drug at a concentration of about 22 mg/kg dry weight (about 0.1 g/kg wet weight). That might not seem like a high concentration, but ivermectin is a poison. If it were available in an aerosol spray to attack house flies or bed bugs (which it also kills), it would probably be at a similar dosage. In other words, ivermectin creates poisonous dung. The chemical is still detectable in cow dung 6 weeks after this has been dropped (Sutton *et al.* 2013). It seems pretty obvious what this will mean for dung-feeding insects.

It wasn't long before entomologists started to worry about the field impact of ivermectin on dung insects. French entomologist Jean-Paul Lumaret (1986) is generally regarded as being the whistle-blower. Thirty years later and there are some startling statistics. Somewhere in the region of 62–98% of the treatment passes unaltered into the faeces, where it remains for weeks. Cows treated with 0.2 mg/kg subcutaneous injection produced dung which was toxic to that saviour of the western Australian meadowlands, *Onthophagus gazella*, for 7 days after administration. Poor thing. Excretion of only 1 µg (that's one-millionth of a gram) per kilogram of dung was toxic to the larvae of the yellow dung fly, *Scathophaga stercoraria*. No more participating in undergraduate field measurements of mate guarding for them. Fly larvae are particularly susceptible to pesticides. The bush fly and some of its relatives might have given dung-breeding flies a bad name, but many of the smaller, more discrete species are an integral part of the nutrient recycling community of the dung, and part of the huge cohort of important pollinators, presently under the political spotlight. Dosing up sheep has a similar effect (Beynon 2012); timings and concentrations may vary, but the final output still contains the toxins.[3]

In field experiments, adult dung beetles were not immediately killed, but they suffered what are euphemistically called 'sublethal effects'; their behaviour and that of their larvae is significantly altered. Egg laying may stop, even though ovary and testes size are normal. Development of the larval stage is delayed; whether this is a slowing down of feeding or a slowing of the metabolism to do with nutrition and growth is not yet clear. Whatever, longer larval periods expose them to extended danger from predators and disease. Sometimes the larvae just stop; effectively this brings slow lingering deaths for the maggots.

In the wild, poisons do not just kill straight away. OK, if an animal dies it is 100% dead, but if it gets a lower dose its physiology can still be damaged; it can become disoriented, it can be less effective at finding food, or a mate, or moving to avoiding a predator. In the complex nesting behaviour seen in dung beetles, it doesn't take much chemical to interfere with the business of tunnelling, rolling, brood ball construction, guarding or tending. Ironically, the death of the dung beetles and their maggots might actually be exacerbating the parasite situation, since removing the dung, fluke eggs and all, into the soil removes them from the infection cycle (Nichols and Gómez 2014).

Some people might baulk at the idea of me suggesting an ivermectin-affected beetle can be 25% or 50% even 95% dead, but in medicine this is the concept of morbidity. A patient doesn't need to be either alive or dead for the doctor to act accordingly. If they are feeling a bit poorly, or are at death's door, they are on opposing ends of this scale. Translating notions of human well-being into the life of a beetle is technically and philosophically challenging, I agree,

3 Extrapolating to other animals is fraught with difficulty. Ivermectin-treated reindeer produced dung which was exactly comparable to non-treated controls, in that no beetles or flies were found. This was because the hard dry winter-browse droppings are unattractive, tough nodules. Only the moist summer dung, from eating lush vegetation, contained a significant insect fauna, but this was not part of the experimental cohort (Nilssen *et al.* 1999).

but another way of looking at it is to think of the affected beetle (or any organism for that matter) as physically, immunologically or behaviourally damaged, or weakened. And if we know anything about natural selection it is that the weak ones are the first to go. They don't stand a chance.

Sublethal does not mean the same as 'alive', it just means 'not dead yet'. It is the sublethal effects of neonicotinoid insecticides on bees and other insects that have got conservation organisations in Britain and Europe into a lather. Just because the organism in question does not drop down dead out of the air immediately, does not mean it is safe. Indeed, it is probably doomed, but conveniently we don't have to watch it die. There are farm animal welfare organisations actively quoting sublethal effects as evidence that populations of dung beetles and flies are hardly affected by the use of endectocides. Any specific effects (e.g. on insect larvae) are claimed to be limited and the total effect on the rest of the population is negligible. They continue to promote and reinforce the message that endectocides remain a valuable choice of product for controlling both ecto- and endoparasites of domestic animals. I put this to Darren Mann of the Oxford University Museum of Natural History, Britain's top professional dung-beetler, I'll warrant. His ripe response: 'Anyone who states that endectocides have no effect on the dung fauna and flora is talking shit.' How apposite, I couldn't have put it better myself, Darren.

One of the more insidious aspects of ivermectin is that, in some cases, dung from treated animals is more attractive to dung beetles than droppings coming from untreated control heifers. This means that the poisonous nature of the chemical is all the more dangerous, because the likelihood of insects being exposed to it is higher. There are also fears that dung beetle fortitude in the face of chemical assault means that some species, or individuals, may be more resistant to it, more likely to survive feeding on it, so more likely to be eaten by predators. This could have very real consequences higher up the food chain, as the drug becomes accumulated in non-target organisms. It was, after all, this exact same effect which concentrated DDT in top bird predators, leading to population crashes of eagles, falcons, kites and all those other raptors still struggling to recuperate decades after the insecticide was phased out. As yet we still do not know enough about how ivermectins move through the food web.

Fears have been expressed, dung-beetlers are off doing research, but conservation organisations struggle to make headway in protecting the environment in general and the dung fauna in particular. Many nature reserves, in Britain at least, are managed for plants, or butterflies or birds, by using traditional, but possibly unprofitable, grazing regimes to maintain flowery downlands, water meadows or ancient pasture woodlands. Can you imagine the kerfuffle if it became known that at the same time as protecting Duke of Burgundy fritillaries or sand lizards they were also poisoning *Emus* and *Copris*?

A correspondent, who will remain nameless, understood the dilemma intimately:

> *A long while back I did some work for English Nature looking at land use and livestock grazing regimes at sites in England in relation to* Asilus crabroniformis *[the hornet robber fly]. I spoke to quite a few conservation land owners who essentially said: 'We don't use ivermectins etc. on the cattle we graze our sites with.' Speaking to the graziers, their take on the matter was: 'We know [so-and-so organization] don't allow ivermectins to be used on the cattle when they are on their land, so we just give them a dose before we take them to site.'*

Urgh.

Harking back to Australia for a moment, it is not difficult to imagine what disruption to the nutrient cycling and dung removal process might look like. Retardation of dung decomposition has long been reported (Wall and Strong 1987). If chemicals are killing, weakening or disrupting coprophage communities, this is likely to exacerbate the situation. It's easy to picture a future where dung pats just sit in the fields, inert, desolate, lifeless. We've been here before.

Ivermectins are not alone in being injurious to dung-breeding organisms. Other veterinary chemicals routinely given to farm animals have similar effects. There is a long list, including dichlorvos, pyrethroids and possibly antibiotics – although some chemicals that decay very quickly, or are more readily expelled in the urine, don't seem to have such a long-lasting environmental effect.

Unfortunately 'good husbandry' is almost a legal prerequisite for animal ownership in Britain. This implies keeping stock animals pest free and healthy. During an email appeal to other coleopterists, Malcolm Story offered me this nugget.

> *I've heard it said that use of ivermectins was 'good husbandry' and animal welfare legislation required all owners to practice 'good husbandry', so ivermectin use was effectively mandated by law. Having surfed a few websites, I can't find that statement, but they take every opportunity to insert the phrase 'good husbandry' in descriptions of product use.*

One of the problems for entomologists is that the beetles do their work for free, and if something is free it is easy to think of it as worthless. I'm not sure how they calculated it, but in an effort to give some commercial weight to dung beetle effort, DUMP (Dung Beetle UK Mapping Project) recently announced that the environmental services provided by even our meagre fauna amounted to £367 million a year.

Even the most ardent dung-beetler must accept that farmers need to raise healthy animals for human consumption, and they need to do this without going bankrupt. But it is the routine, unthinking, robotic, prophylactic administration of these pharmaceuticals that is causing the problem. There must be a balance between chemical pest control and chemical-free dung communities. It's just that we haven't got there yet.

MEGAFAUNA AND MICROFAUNA EXTINCTIONS

It's a sad fact of history that whenever humans arrived at a new land, they managed to hunt the large charismatic megafauna to extinction. In Australia 50,000 years ago they eradicated *Diprodon* (a marsupial the size of a hippo), *Procoptodon* (a horse-sized kangaroo) and the marsupial lion *Thylacoleo*. After the last ice age, 15,000–10,000 years ago, humans spreading north into Europe extinguished the mammoth *Mammuthus* and the sabre-toothed tiger *Smilodon* and all the dwarf elephants from the Mediterranean islands of Crete, Cyprus,

Malta, Sardinia and Sicily. At the same time they crossed the Bering Strait, from Chukotka into North America, to erase *Gliptodon* (a giant armadillo-type thing), mastodon (*Mammut*, slightly smaller than a mammoth) and *Megalonyx* (a ground sloth weighing a tonne). The extinctions continued throughout the New World as hungry hunter humans finally made their way down to Tierra del Fuego. We know this from the bones still being found. We can only guess at the dung beetles, which the droppings of these animals once supported, that are now gone too.

In Argentina palaeontologists are still unearthing fossilised dung balls as big as, or bigger than those rolled round Africa today. We don't really know for sure what laid the dung, or what shaped them. There are no suitable large roller beetles left in South America today. The extinction of one set of animals created a cascade that eradicated others too. Just as depletion of prey items will starve the predators, the hunting of the large mammals left nothing for the despondent dung beetles.

There's nothing we can do about what might have been beautiful glossy mastodon mega-roller beetles, or bizarre diprodon dung tunnellers disporting debonair horns. Extinction is forever. But we might have a go at preventing the same thing happening again. There is another megafauna extinction going on today. Elephants, rhinos, gorillas, tigers and pandas are just some of the large mammals now in precipitous decline around the world. Whether they are being hunted for meat and trophies, or being backed into a corner through climate change, habitat destruction and human encroachment, they are at the edge of the abyss. And so too are the dung beetles they feed with their droppings. Already, African dung beetles are feared to be on the decline (Nichols *et al.* 2009), joining the similar misfortunes piling on to the hunched shoulders of European scarabaeids in Italy (Carpaneto *et al.* 2007), Iberia (Lobo 2001), France (Dortel *et al.* 2013) and Britain (Dung Beetle UK Mapping Project). In Australia, too, dung beetles may also be declining, not through competition with all the exotics brought in, but because of a continuing extinction of native mammals (Coggan 2012). Today, world-wide, 12% of all dung beetle species are reckoned to be 'endangered' (on the brink of extinction) or 'vulnerable' (fast heading towards that brink). There are dark times ahead for dung beetles.

This is not just an observation of detached, dispassionate, scientific curiosity. There are real ramifications to human destruction of the ecosystem. The complex ecology of the world, very well exemplified by Australia's dung-based tribulations, is easily upset. It is quite likely that had things gone slightly differently back in Australia's prehistory, if the hunting had not been quite so thorough, that the whole cow dung debacle could have been avoided. Imagine a world where giant diprodons and marsupial tigers still patrolled the open savannahs, leaving their droppings as they went. The large rollers and hearty tunnellers evolved to take advantage of this copious manna would, perhaps, have been better equipped to deal with the heavy cow pats when they started to drop in 1788.

I've admitted from the very start that I'm biased about beetles. Beetles are very important, not just for their handsome lustrous bodies and perky athletic forms, but because of their huge worldwide diversity and their amenability to scientific study. We can't look at everything, everywhere, all the time, but just looking at beetles gives us a window into the workings and the failings of the ecosystem of which we, humans, are a part. They are our barometer of environmental health, our measure of ecological resilience, our early warning system of impending disaster in the biosphere. We need to pay more attention to beetles, and dung beetles in particular are well worth a closer look.

When Anderson and Coe famously reported 16,000 beetles dispersing an elephant dropping in under 2 hours, or Heinrich and Bartholomew raged enthusiastically about a cupful of dung attracting 3,800 beetles in just 15 minutes, they were trying to express in numbers just how powerful and awe-inspiring these diminutive creatures could be. The same power and awe was felt by a ten-year-old boy when first a dumbledor pushed its way out through clenched fingers and flew off into that chalk downland evening long ago. It would be a terrible and heavy shame on us all if these tales became the stuff of mere legend.

CHAPTER 11

DUNG TYPES – AN IDENTIFICATION GUIDE

Dung varies greatly, what goes in one end really does affect what comes out the other, even for the same animal at different times of year, even on different days. This guide is meant as a very general introduction to scats and droppings, and its main purpose is to show the variety in form and how this reflects an animal's diet, its behaviour and its life history.

I've always been rather cavalier in my examination of pats and stools, burrowing in with carefree abandon using a knife, trowel, a handy stick or sometimes even my bare hands. As has been discussed, excrement is full of bacteria, both benign and dangerous, as well as other unsavoury organisms and diseases. As far as I know I have never suffered any ill effects from my ridiculously unguarded behaviour, but this is the modern world, and it would be negligent of me to suggest anything other than extreme caution to the inexperienced reader.

So, avoid touching the dung with your fingers; use a sharp stick or stout twig to turn over any pellets or droppings; wash your hands thoroughly when you have finished, and certainly before going off to eat. Even I always made sure I used a different knife when it came to slicing up the apple in my packed lunch.

Cow, *Bos taurus* – pat, pad, pie, cake or muffins (when dried), buffalo chips (North America)

Size: 20–50 cm diameter, 20–60 mm deep.

Description: Semi-liquid, the consistency of wallpaper paste, forming a roughly round or oval flattened dollop, splattering if dropped onto hard surface. Nominally brown, but usually with deep olive greenish tinge. Can appear bright orange or yellow in unweaned calves.

Location: Fields, meadows, droves, cow-sheds – wherever cows graze or are housed.

Ecological notes: One of the most water-rich of terrestrial dungs, forming a dense smothering pool on the grass where it falls. Outer surface soon forms a skin, then a crust in hot dry weather. Short fibres of undigested plant material clearly visible inside. Caused ecological chaos in Australia where native dung-recyclers could not cope with it, and it remained unbroken down, removing large areas of grass from useful grazing production until introduction of African and Asian species more adapted to the wetter bovine excrement. Often regarded as the default country manure smell. Buffalo and bison dung similar, slightly less liquid, forming ridged or folded pat.

Horse, *Equus ferus caballus* – road apples (originally North America)

Size: 25–45 cm across, 10–20 cm high.

Description: A rough, loose heap of conjoined, but individual boluses forming a pile, or dropped in a line as the horse walks. Mid to pale brown, sometimes beige or yellowish, with long, undigested grass or hay strands clearly visible. Slightly moist when fresh, but easily drying to crumbly matted-straw consistency.

Location: Fields, meadows, stables, roads and bridleways. Horses noted for ability to drop dung without breaking step. Horse-drawn tours of many cities now required to use dung-catcher hammock-like sling, draped under the beast's hindquarters.

Ecological notes: Highly attractive to a wide variety of dung beetles and flies, and the entomologist's top choice for a day spend dunging. Fruity, fermented, almost sweet smell makes it a popular choice for garden or allotment manuring. Donkey and zebra dung similar.

Sheep, *Ovis aries* – buttons, treddles or trottles (dags are small pellets entangled in the wool near the tail)

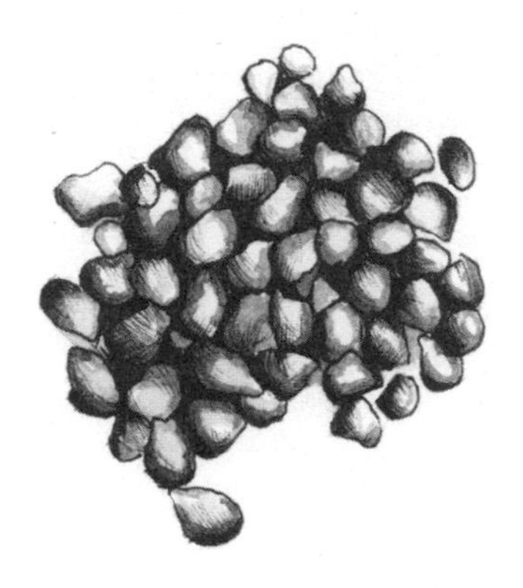

Size: 5–15 cm long, 3–6 mm diameter.

Description: Tight clusters of smooth, dark pellets, each 1–2 cm across, compressed into a sausage-shaped stool that often breaks on impact with the ground. Light or dark brown to almost black. Smooth and glossy when fresh, drying to hard nuggets.

Location: Fields, downs, hills and moors, wherever sheep graze.

Ecological notes: Hard, dark, dense pellets, with only inoffensive scent. Pellets become compacted into a dropping in the rectum, but the fracture lines remain when the dung drops and these provide initial entry points for the usually small dung beetles that push inside. Since sheep are well adapted to hill living (rather than the lowland grazing reserved for cows), the dung beetles found in sheep dung often happen to be specialist limestone downland, moorland or even boreo-montane species.

Moose (called 'elk' in Europe), *Alces alces*

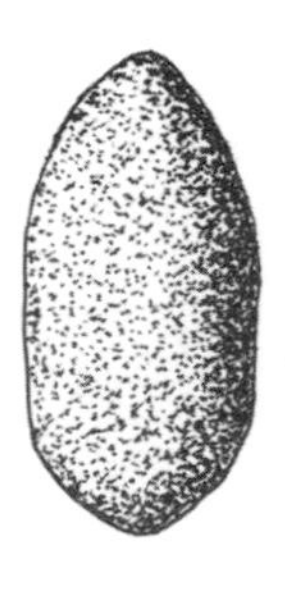

Size: 15–20 mm across, 20–30 mm long.

Description: Really just a very large deer species, so droppings also very similar. Two types of pellet: one type is larger, short, sub-spherical, with one end pointed the other flattened or slightly concave; the other type is more evenly rounded, longer oval with only a vague point on one side. In winter the pellets are pale, dry and firm, but in summer they are moist, sticky, deeper coloured, almost black.

Location: Woods and mountainsides.

Ecological notes: Some of the older books suggest the two types of pellet are produced by the two sexes, the large, short, spherical ones by the bulls and the small, longer, oblong pellets by the cows, but there are many intermediates and no true sex bias in droppings occurs. However, hunters claim that bulls often drop dung in a close group when standing, while cows scatter theirs as they walk.

Reindeer (caribou), *Rangifer tarandus* (papana in Lapland)

Size: individual pellets 8–15 mm across like those of moose (winter) or a pat 10–15 cm across (summer), shown here.

Description: Hard, round or roughly egg-shaped pellets usually scattered in loose archipelago. Brown of varying shades, sometimes greyish.

Location: Mountain and moorland slopes, where reindeer graze.

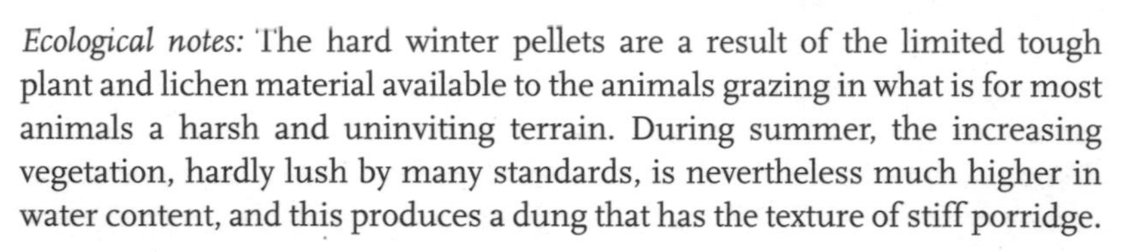

Ecological notes: The hard winter pellets are a result of the limited tough plant and lichen material available to the animals grazing in what is for most animals a harsh and uninviting terrain. During summer, the increasing vegetation, hardly lush by many standards, is nevertheless much higher in water content, and this produces a dung that has the texture of stiff porridge.

Deer, various species – fumes, fewmets, fumets, fewmeshings, cotyings

Size: 10–35 mm across.

Description: Sub-spherical pellets, sometimes slightly cylindrical, rounded, but usually with a point at one end, acorn-like. Dropped singly, or in compacted clumps of 10–20, which usually break apart on hitting the ground. Normally black and smooth on exterior (especially in summer when they are soft and sticky), but yellow-green or brownish inside.

Location: In deer parks scattered on fields and meadows, but otherwise secretively in woods, on rides and paths, clearings or places where the deer have hidden up.

Ecological notes: Size is a useful indicator of deer species: wapiti (called 'elk' in North America), *Cervus canadensis*, 13–18 mm across and 20–25 mm long; red deer, *Cervus elephas*, 12–15 mm across and 20–25 mm long; fallow deer, *Dama dama*, 8–12 mm across and 10–15 mm long; roe deer, *Capreolus capreolus*, 7–10 mm across and 10–14 mm long; muntjac, *Muntiacus reevesi*, 7–10 mm across. Dung beetles with close deer dung association are not choosing the droppings because of scent or nutritional make-up, and are more likely to be limited by being shade specialists, as compared to open meadowland specialists in cow and horse dung.

Giraffe, *Giraffa camelopardalis*

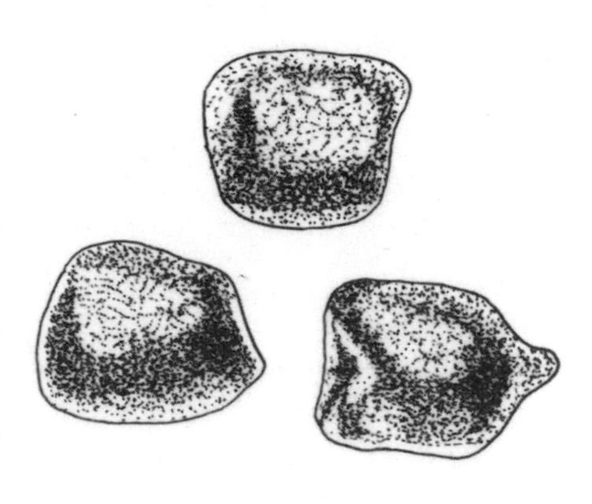

Size: 2–3 cm across.

Description: Small, barrel-shaped pellets, nominally cylindrical, usually with one flattened end, one end slightly pointed. Dark, smooth, glossy brown, drying to paler straw colour.

Location: Savannahs of southern Africa.

Ecological notes: Despite the very large size of the animal, the pellets are remarkably small. Although many pellets are extruded in a single compacted dropping, they scatter, often widely, when they hit the hard ground because of the long drop. As with other grazers, giraffes excrete copious amounts of dung, around 70 kg/day.

Camel, *Camelus dromedarius* and *C. bactrianus*

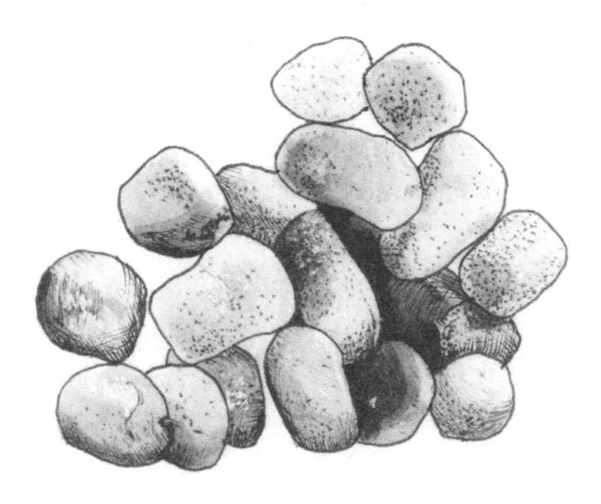

Size: 3–6 cm across.

Description: Large rounded oval pellets, roughly the size of a chicken egg. Dark to mid brown.

Location: Scattered wherever the camels walk. Nominally in dry places, where camels abound – deserts and rocky landscapes – but plenty of other pasture-like landscapes, since they are domesticated in many parts of the planet.

Ecological notes: Adaptation to dry-landscape living means that camels extract as much moisture as possible from the gut contents before it is voided. This gives extremely dry dung, which can be burned fresh from the camel. This is an important fuel in north Africa, Arabia and Central Asia. Eating fresh camel droppings reputed by bedouins to cure dysentery, by alteration of the bacteria in the human gut. Don't do this at home, folks.

Elephant, *Loxodonta africana* (African) and *Elephas maximus* (Indian)

Size: 30–50 cm across, 20–30 cm high, weighing 1.5–3.5 kg.

Description: Large pile of rounded boluses. Bright olive yellow when fresh, with a reputedly rather strange and pleasant smell. Becomes paler as it dries over several days.

Location: Savannah, grassy plains, forest and desert.

Ecological notes: Freshness can be tested by thrusting a hand deep into the pile, warmth indicates a recent drop. The dung is very fibrous, sometimes little more than chewed herbage and is attractive to other herbivores such as deer, and to monkeys and francolins (small pheasant-like birds) which pick out the seeds. Huge numbers of beetles have been recorded from single droppings.

Rabbit, *Oryctolagus cuniculus* – crottels, croteys, crotisings, etc.

Size: 8–9 mm diameter, occasionally 10 mm.

Description: Spherical, mid to dark brown, becoming lighter as they dry with time. Remains of plant stems and fibres clearly visible on the surface. Look as if they have been kneaded from damp cigarette tobacco.

Location: Clustered in dense latrines near the centre of the rabbit colony. Often on raised feature such as grass tussock, mole or ant hill, tree stump.

Ecological notes: Although not strongly smelling to the human nose, latrines are thought to act as territorial scent marking. Pellets at a latrine are often darker because they have also been splashed with urine. The night faeces, produced after first digestive transit, are never found, as these are taken into the mouth straight from the vent while the animal is deep inside the safety of the burrow. In North America cottontails (*Sylvilagus* species) are very similar, but the closely related pikas (*Ochotona* species) have pellets 3–4 mm across.

Hare, *Lepus europaeus* – crottels, croties, crotishings, etc.

Size: 12–18 mm diameter.

Description: Spherical, slightly flattened, very like large rabbit droppings. Pale, sometimes yellowish brown in winter; darker brown, with greenish tints, almost black, in summer.

Location: Scattered widely, sometimes in small groups near the form (the meagre nest scrape in the soil), usually around feeding areas, but not in distinct latrine areas.

Ecological notes: Apparently smelling of damp digestive biscuits, with a hint of mown hay. North American arctic hares, snowshoe hares and jackrabbits are very similar.

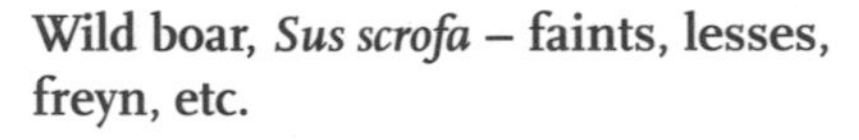

Wild boar, *Sus scrofa* – faints, lesses, freyn, etc.

Size: 10–15 cm long, 5–8 cm diameter.

Description: Large, dark, irregular or sausage-shaped droppings made up of many small oblong pellets compacted together. May break up into shorter fragments as they dry. Brown, dark grey or black.

Location: Woods, wooded hills and mountainsides.

Ecological notes: Dung of domestic pig is similar, but varies more in colour and texture depending on its feedstuffs. Traditionally a pig, kept in a sty, was fed swill from domestic leftovers. This omnivory can lead to rather human-like dung, with associated looks and smells, causing more than the usual upset at rural scents from neighbours of pig farms. The reintroduction, or accidental escape, of wild boar back into British woods after a gap of about 700 years promises to show some interesting results if ever the dung beetle fauna of their droppings is reported.

Fox, *Vulpes vulpes* – fuants, billets, billetings, scumber, waggyings

Size: 5–10 cm long, 2–3 cm diameter

Description: Sausage shaped, cylindrical, with spirally twisted point at one end. Varying from black to pale grey, fibrous to squishy depending on what has been eaten.

Location: Fields, parks, gardens, roads. Often deposited in prominent position on patch of bare ground, doormat, low wall, garden furniture or toys left outside. This is obviously a territorial display, combined with scent marking.

Ecological notes: Typical dropping oily grey, but can be shiny black, older dung pale grey as it dries out or white if containing crunched bones. In late summer droppings are shining knobbly black with red and purple highlights after eating berries. Highly fragrant and, to my mind, the most offensive of dungs.

Dog, *Canis familiaris*

Size: 5–35 cm long, 1–8 cm diameter.

Description: Cylindrical sausage-shaped. Virtually any colour from pale grey, almost white, through yellow, orange, red and brown to near black.

Location: Pavements, gardens, parks, paths, verges, anywhere the damn animal cares to drop it.

Ecological notes: Huge variety in shape and size because of vast genetic variation in domestic breeds. Colour range of dung from diverse range of mostly canned foods available for pet-owners, supplemented with whatever human leftovers, snacks and treats are given. Strongly smelling, noisome and offensive. Wolf (and coyote) dung is very similar, but with more texture, containing fur, feathers, bones and other non-digested material; called freyn, lesses, faints, fuants, etc. The main genus of puff-ball fungi is *Lycoperdon*, reputedly from the Greek λυκος (*lukos*) for wolf and περδομαι (*perdomai*) meaning to 'break wind behind', because they (or truffles depending on your classical translation) were thought to grow from wolf dung.

Cat, *Felis catus* – scat

Size: 3–7 cm long, 1–2 cm diameter.

Description: Cylindrical smooth sausage-shaped, usually with narrowed point at one end, sometimes coiled. Brown, grey or black.

Location: In gardens, usually recently dug-over areas, neatly sieved and raked ready for planting; will bury it under a layer of soil if they can be bothered. Also on gravel paths, and in children's sandpits.

Ecological notes: Stronger smelling, because of more meat-based feeds, than dog dung, despite most food being mass-produced and processed, provided by owners from cans and sachets. Dung from big cats is similar, but larger – lion (*Panthera leo*), 10–20 cm; leopard (*P. pardus*), jaguar (*P. onca*), puma (*Puma concolor*), all about 8–15 cm; cheetah (*Acinonyx jubatus*), 18–25 cm; lynx (*Lynx* species), 20–25 cm; etc. – and containing more fur, feathers, bones, etc. A recent kill, with high blood and meat intake will produce a very black, strong-smelling scat. In northern and montane areas, a lynx will bury its scat under snow.

Spotted hyena, *Crocuta crocuta*

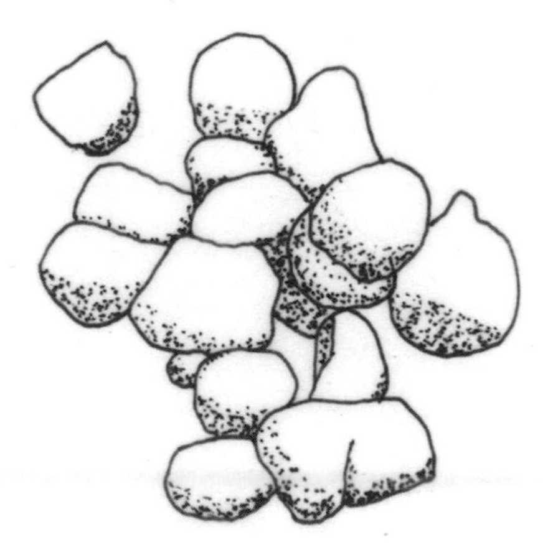

Size: 5–7 cm across.

Description: A cluster of roughly round boluses, greenish when fresh but turning pure white with age. Sometimes smooth, sometimes coarsely fibrous, packed with hair.

Location: Savannahs of southern Africa, usually in a conspicuous latrine in an open area.

Ecological notes: The whiteness that appears is because of the bone content of the food – hyenas have immensely powerful jaws and teeth to grind even quite large bones to powder. Droppings of the smaller brown hyena, *Hyaena brunnea*, are similar, but narrower, roughly 4–5 cm across.

Badger, *Meles meles* – faints, fuants, werderobe

Size: 5–8 cm long, 1–2 cm diameter.

Description: Cylindrical sausage-shaped, slightly pointed at one end. Black or grey or brown. Surface uneven or rough, sometimes oily, semi-liquid.

Location: Near the entrance to the sett, in a small oblong hole about 10 cm long and about 6 cm deep, dug with its forepaws, but not covered over with soil after use. A series of such pits form a latrine, and each hole may be used several times.

Ecological notes: The latrine is often quite obvious near the home diggings of the sett and is thought to be a territorial marker to indicate that the burrow is occupied. Dung is sticky and smelly when fresh.

Otter, *Lutra,* and other genera – spraints, spraits

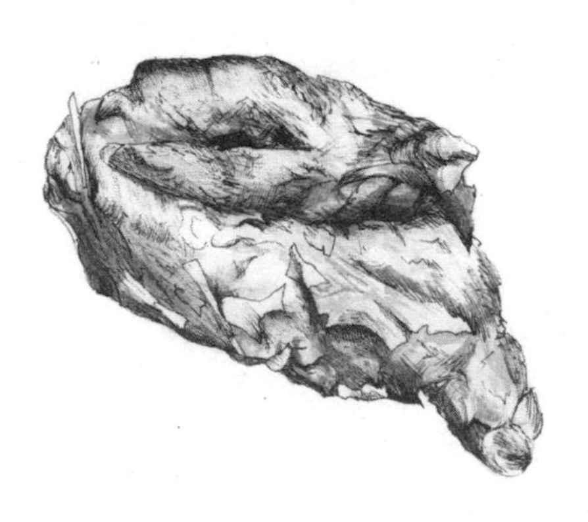

Size: 3–12 cm long, 10–15 mm diameter.

Description: Black, slimy and tar-like when fresh, drying to a pale grey and becoming crumbly. Persistent oily smell, sometimes likened to violets. Contains fish bones, scales, crustacean carapaces and sometimes fur or feathers.

Location: Usually elevated areas near the river bank of lake shore, especially small peninsulas jutting out into the waterbody; often on small hillock of sand or vegetation that the otter has scraped together.

Ecological notes: Distinct spraint scent may be used as a territorial marker, this would explain prominent depositions. Otters also leave smears of thick anal jelly, variously white, black, orange or brown, thought to act as a lubricant to allow easier passage of faeces which are thick with sharp fish bones and scales.

Pine marten, *Martes martes* – dirt, faints, fuants

Size: 4–12 cm long, 8–15 mm diameter.

Description: Long, thin, twisted, dark, tapering to long, pointed strand at one end. Contains visible remains of bone, fur, feathers and plant material. Scent often regarded as pleasant, musky or fruity.

Location: Woodlands, deposited prominently on tree stump, log, rock or wood pile.

Ecological notes: Yellowish droppings indicate eggs have been eaten; blue or red that it has been eating berries. Similar but smaller beech marten (*Martes foina*) in more open country, with inclination to nest in lofts and outhouses; its excrement claimed to be evil-smelling, but authorities disagree on this subjective measure. Polecat (*Mustela putorius*) and mink (*Mustela lutreola* and *Neovison vison*) also similar. Mink often with looser stools after feeding on fish or frogs.

Weasel, *Mustela nivalis*

Size: 2–3 cm cm long, 2–5 mm diameter.

Description: Long, thin, twisted, rope-like, dark, often black, but sometimes grey, containing fur, feather and bone fragments. With a distinctive musty smell.

Location: On raised or prominent position such as grass tussock, anthill, rock or tree stump; a territorial marking. Old and new droppings sometimes found together at the same drop point.

Ecological notes: Not as strong smelling as marten droppings. Those of stoat, *Mustela ermine*, and polecat, *M. putorius*, are similar but larger, 6–8 cm long, 8–15 mm diameter.

Hedgehog, *Erinaceus europaeus*

Size: 3–5 cm long, 8–10 mm thick.

Description: Shiny, black, cylindrical, vaguely lobed, usually pointed at one end.

Location: Garden lawns, parks, paths and pavements

Ecological notes: The remains of hard insect shells are often visible, and these partly give an idea of what the animal has been eating. Fur or feathers in a small, thin, dull, twisted dropping, resembling that of stoat or weasel, indicate it has eaten a mouse or nestling. Faeces are sloppy when earthworms have been devoured.

Bat, Chiroptera, various species

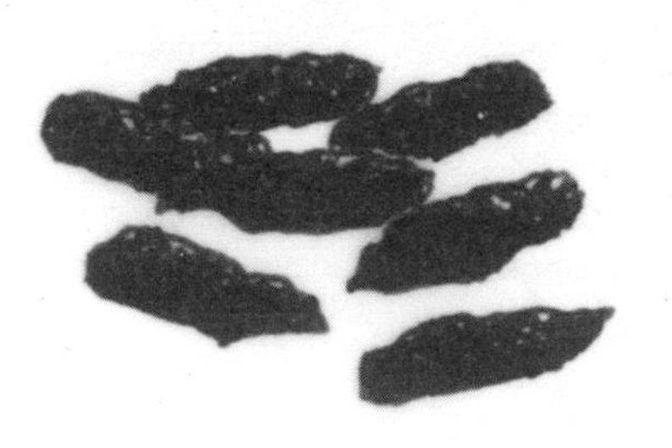

Size: mostly small, up to 6–8 mm long, 2–3 mm diameter.

Description: Small, dark, cylindrical with one end slightly more pointed than the other. Dark brown to black. Rougher and more porous than mouse or rat droppings.

Location: Scattered on the floor at the entrance to roosts, so all over the patio or veranda if nesting up under the house eaves. Also in the loft, or in cellars, where they may be roosting, and if undisturbed these may accumulate into piles and drifts. Throughout the world various cave-roosting species occur in huge numbers and their guano lies many metres thick on the floors of the caverns.

Ecological notes: Insectivorous species in Eurasia and North America have droppings made up of insect remains, the tough chitinous shells being broken, but often undigested. Vampire bats (three scarce South and Central American species) live solely on mammalian blood, so their excrement is like a sticky black pungent tar that tends to gather under their roosts. Fruit bats (flying foxes) of tropical Asia and Australia are eaters of fruit-juice and pulp. Their droppings are larger, but mushier, and contain seeds rather than insect remains. They mostly suck fruit juices (smaller species nectar) but do not eat the pits or stones, instead they spit them out; these 'spats' can cause consternation to house owners if they rain down too heavily onto roofs, getting lodged and blocking gutters.

House mouse, *Mus musculus*

Size: About 6 mm long, 2.0–2.5 mm diameter.

Description: Small, dark, irregularly cylindrical pellets, one or both ends slightly pointed. Relatively solid, fibrous, moist, smearing easily between finger and thumb when fresh; harder, crumbly when old and dry.

Location: On floors in barns and granaries, inside cupboards and on shelves in the kitchen. Scattered seemingly at random as they forage. 50–80 droppings per night are quoted.

Ecological notes: Seemingly innocuous, non-smelly and dry pellets, but implicated in disease spread, and unacceptable in foodstuffs; a sign usually reacted upon swiftly by the householder. The 'smell' of mice, recognised by those familiar with it comes from the urine, rather than the faeces. Droppings of wood mouse, *Apodemus sylvaticus,* are similar in size, but are paler, more fibrous.

Black rat, *Rattus rattus*

Size: About 10 mm long, 2–3 mm diameter; of brown rat, *Rattus norvegicus*, about 17 mm long, 6 mm diameter.

Description: Oval-cylindrical, slightly curved with a point at one end, dark brown. Those of the brown rat are larger, fatter, more oval, like olive stones.

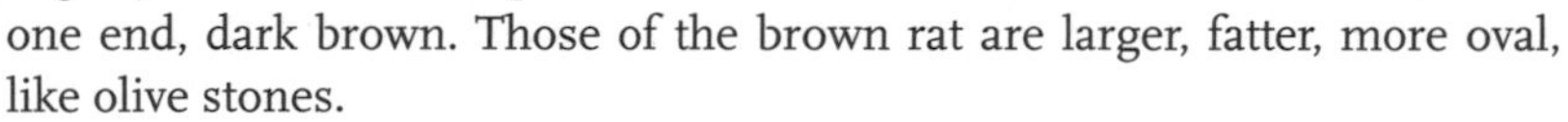

Location: Scattered wherever the animal forages, building up into broad spread of droppings over the floor if an infestation is left unchecked. Brown rats have more of a tendency to use latrine sites and in similar situations will defecate in the corners. 40–50 pellets per day are quoted.

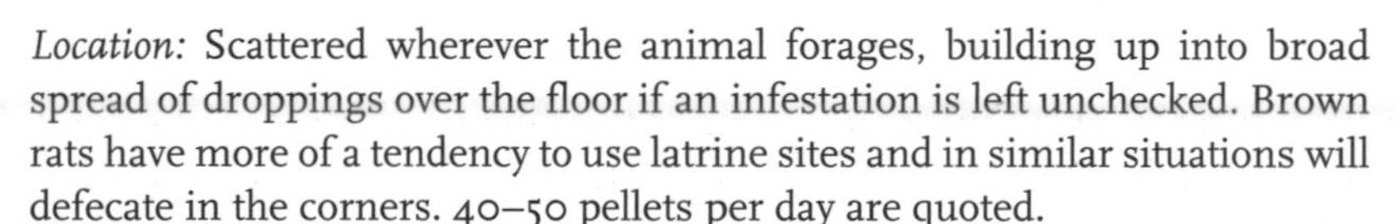

Ecological notes: As with mice, droppings are often the first sign of an infestation, long before a live animal is spotted. Rat smell is from the urine. Both urine and faeces implicated in human disease spread.

Water vole, *Arvicola amphibius*

Size: 7–10 mm long, 3–4 mm diameter.

Description: Smooth, cylindrical, with rounded ends – like short fat cigars. Dark brown, greenish when fresh, soft, the texture of putty and containing very fine plant shreds. No strong scent.

Location: River and stream banks, pond sides, gathered in latrine heaps, often trodden flat by later activity.

Ecological notes: Because they are so nervous and secretive, water voles are hard to observe. In the UK, they were found to be declining drastically after non-native mink escaped from fur farms in the 1960s, and set up large feral colonies along the same rivers and streams. Counting water vole latrines is now a standard survey method, and recognition (along with feeding signs) is taught to any surveyors recruited.

Common shrew, *Sorex araneus*

Size: 2–4 mm long, 1 mm diameter.

Description: Small, dark, cylindrical packages, often twisted into a point at one end. Dark grey or black when fresh, fading to brown when older.

Location: In small piles in runs through grass root thatch, near nest entrances, under logs and stones.

Ecological notes: Brittle and crumbly as they contain many insect exoskeletons. Water vole droppings contain distinct pale grey or white crustacean fragments. Tree shrews on Borneo (*Tupaia montana*) use giant pitcher plants (*Nepenthes rajah*) as convenient toilets. The plants derive nutrition from the gifted faeces and in return they provide nectar to their shrew visitors.

Red squirrel, *Sciurus vulgaris*

Size: 4–6 mm across.

Description: Varying in shape from broad oval to sausage shaped. Smaller than rabbit pellets, but shorter and fatter than rat droppings. Brown or grey, smooth on the outside, but fibrous within, often flatted slightly at one end, pointed at the other.

Location: On logs, tree stumps, fence posts, upper surface of tree branches and other places where the animal has been sitting to feed.

Ecological notes: Grey squirrel (*Sciurus carolinensis*) has similar droppings, perhaps slightly larger, to 10 mm, as the animal itself is bigger. The grey will sometimes nest in old lofts and will scatter its droppings across the floor. Chipmunk (*Tamias* and other genera) have similar, but slightly smaller scats.

Porcupine, *Erethizon dorsatum*

Size: 15–35 mm long, 10–15 mm diameter.

Description: Irregular smooth or rough pellets, very like those of deer, but less regular in shape and size. Longer droppings slightly curved.

Location: On paths, or scattered under trees. During the winter an animal spends a lot of time in a single tree, and a heap of pellets accumulates at the base of the trunk. Brown to almost black. Contents fibrous. Similar latrines occur in or near caves, rock hollows or tree holes in which the animal is sheltering.

Ecological notes: Texture and shape depends on foodstuffs. Winter droppings tend to be more irregular in shape, and uneven on the surface, as the porcupine eats bark and twigs; summer droppings are smoother and moister, from the more succulent vegetation in the diet. A series of pellets can be joined together, like a necklace, linked by strands of vegetation – stems or grass stalks.

Coypu, *Myocastor coypus*

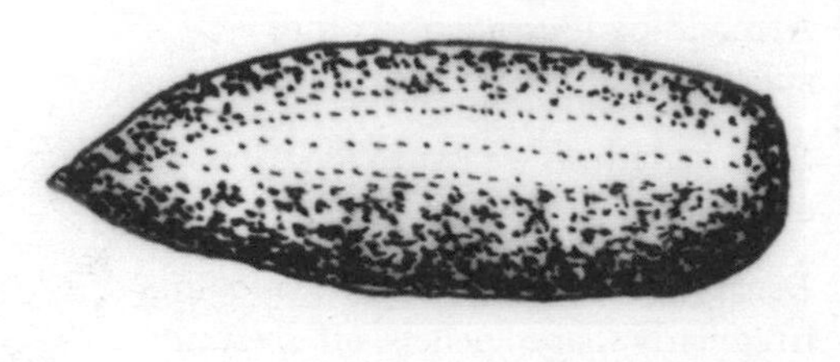

Size: 2–4 cm long, 1 cm diameter.

Description: Bullet shaped, cylindrical, blunt at one end, pointed at the other. Smooth, but with series of distinct wrinkles, sometimes forming parallel grooves and ridges along its length. Dark to mid brown, sometimes greenish or black.

Location: River and stream banks, marshy places, sometimes floating on the water near the shoreline.

Ecological notes: This large rodent is native to South America, but feral colonies are established in North America and in Europe following escapes from fur farms. It was successfully eradicated from the UK in the 1980s after burrowing damage to river and dyke banks. Monitoring for its droppings has confirmed its continued absence.

Eurasian beaver, *Castor fiber* (and *C. canadensis* in North America)

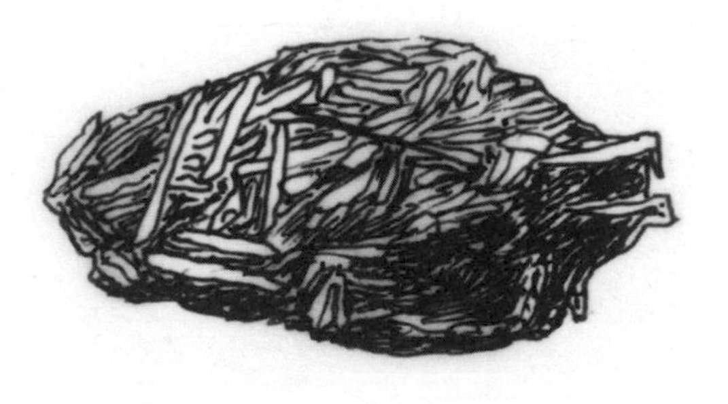

Size: 2–4 cm long, 2 cm thick.

Description: Oval, rounded, vague point at one end. Dark brown, becoming paler as they dry with age. Contain very coarse plant material, often resembling sawdust, which indeed they may also contain.

Location: River and stream banks, marshy places, but usually found floating in the water, although they do not survive long and soon disintegrate. Most often found in the early morning at the water's edge.

Ecological notes: Beavers reingest their own soft greenish-brown 'night faeces' like rabbits, to gain extra nutrients from their highly fibrous diets. This is never found because they take the pellet directly into the mouth from the vent, in the safety of their nests.

Armadillos, *Dasypus* and other species

Size: Length to 45 mm, 10–20 mm diameter.

Description: Relatively smooth, irregularly shaped pellets, often round marbles. Sometimes compressed together into larger sausage-shaped droppings.

Location: Sometimes gathered in small latrines, especially in dry locations, but some are also buried.

Ecological notes: Armadillos are insect-eaters, using their strong clawed front legs to dig them up. They end up eating a lot of soil with their prey, and the droppings are often mostly made of clay. Armadillos are regarded with fond admiration in the southern USA and dark chocolate praline drops are now available branded as 'Texas Armadillo Droppings'. Yum.

Aardvark, *Orycteropus afer*

Size: 3–4 cm long, 2 cm diameter.

Description: Oval oblong nuggets, smoothly rounded, but sometimes uneven or irregular.

Location: Dry savannahs in southern Africa

Ecological notes: Almost completely made up of sand, the animal having more or less fully digested its food items – many thousands of termites at a sitting – but passing the soil material which it eats incidentally when it is feasting. The aardvark cucumber is the only fruit eaten by the animal, for its water content; the seeds are then spread in the aardvark's droppings, which it helpfully buries, ready for the germination. Aardvarks will also dig up and eat dung beetle pupae.

Kangaroos and wallabies, *Macropus* and other genera

Size: 8–20 mm diameter.

Description: Rounded, hard, dry nuggets, darker brown outside, pale and fibrous inside. Very similar to hare or rabbit droppings.

Location: In feeding areas, often in groups of four to eight.

Ecological notes: Australian dung beetles adapted to dry marsupial dung famously could not cope with high water content of cow and horse dung, which started to smother meadowland after the introduction of these domestic animals in the 18th century. Apparently makes good tinder when lighting camp fires.

Brown or grizzly bear, *Ursus arctos*

Size: 10–25 cm long, 6 cm thick. Black bear, *Ursus americanus*, 10–20 cm long, 4–5 cm thick.

Description: Extruded as cylinder, but breaks up into pieces in the pile. Formed of smaller pellets, but these become compacted into a stool. Nominally dark brown, but paler in summer when they contain a large quantity of hair, fur, bone, insects and plant material. During the salmon run season, in June, scats have a strong fish smell and contain many small fish bones. In autumn runnier, black, with hints of purple or red when the animal has been eating berries.

Location: In the woods, obviously. Forests, mountain woodlands, dropped wherever the animal is feeding.

Ecological notes: Since bears are highly unfussy eaters, their droppings can vary tremendously. During salmon spawning season the dung smells strongly of fish, and is full of fish bones and scales. Scat of polar bears (*Ursus maritumus*) is being used to discover what they eat as the polar ice retreat affects their available prey items – less seal, more caribou, seagulls, goose and goose eggs.

Wombat, *Vombatus ursinus*

Size: 2 cm across.

Description: Hard, dry, mid to dark brownish black with green or ochre tinges. Cubic. Not sharp-edged like a box, but flat-sided or gently convex on the normally six faces, making them look very like miniature bread loaves.

Location: In latrine areas, to mark territory, 6–8 pellets dropped at a time, up to 100 a day recorded.

Ecological notes: Quite how the pellets become six-sided is still a bit of a mystery. Wombat digestion is a long drawn-out process, taking 14–18 days from eating to defecation, allowing maximum nutrition to be obtained and optimum water reabsorption for a mammal living in arid conditions. Spherical boluses from the small intestine become tight packed in the large intestine causing the sides to flatten where they butt up against each other before they are eventually voided.

Human, *Homo sapiens* – stool

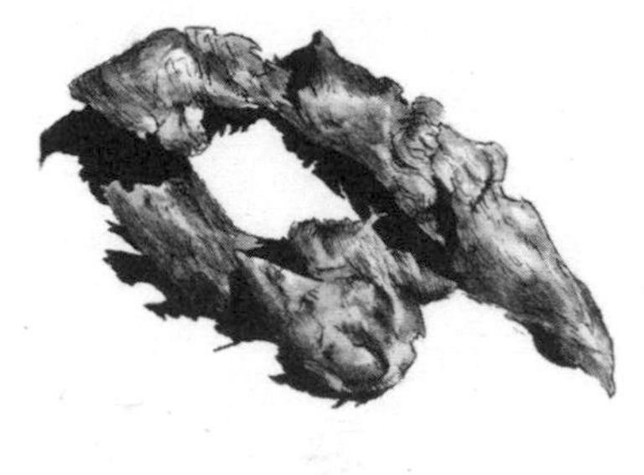

Size: Usual length 10–30 cm long, 20–50 mm diameter.

Description: Typically deep brown, moist, flexible, tubular, smooth or gently undulating surface, though varies from small, hard, separated dark, almost black, pellets (constipation) to yellow or orange watery or frothy liquor (diarrhoea). Weight approximately 500 g daily.

Location: In developed countries mostly flushed away in watery waste from domestic lavatories, deposited in chemical toilets or composting systems, unseen except by the few sewer cleaners or workers in sewage treatment plants. However, sometimes deposited 'in the wild' during countryside walks where no public conveniences are convenient. In some nomad cultures (including military) they are buried individually.

Ecological notes: Produced in vast quantities from the dense urban populations of towns and cities so prompt removal to suitable treatment sites becomes an important health and social consideration. The human omnivore diet makes any available dung attractive to a wide range of dung-recycling insects, making it the bait of choice for entomologists studying dung fauna.

Whales, various species

Size: Difficult to measure, probably scores or hundreds of kilograms at a time.

Description: A cloud of semi-liquid particles, like a giant lava lamp has been emptied, released as the whale is swimming, brownish or reddish. That of the blue whale is quite firm, and fishy.

Location: Wherever the whale pleases.

Ecological notes: Whale faeces are thought to play a vital role in nutrient cycling in the ocean. Many whales feed in the depths, but defecate near the surface, reversing the usual nutrient path from sunlight on algae near the surface, through algal feeders, predators, scavengers and death eventually leading to nutrients falling down the water column into the abyss. Some whales (sperm whales, etc.) produce ambergris, a waxy secretion that may protect the digestive tract against sharp squid beaks. When this is ejected, it floats and is washed up on shore. Muskily scented, it is a valuable commodity used in the perfume industry.

Goose, various genera – gaeces has been suggested

Size: 5–10 cm.

Description: Thick, cylindrical, coiled or curled. Mushy green-grey from digested grass, usually with white uric acid patches predominating at one end.

Location: Often messing paths and lawns near lakes in parks and ornamental gardens. Easily trodden flat by the constant comings and goings of the birds. Heaps of droppings occur, along with lots of feathers, when a birds remains in one small area to moult.

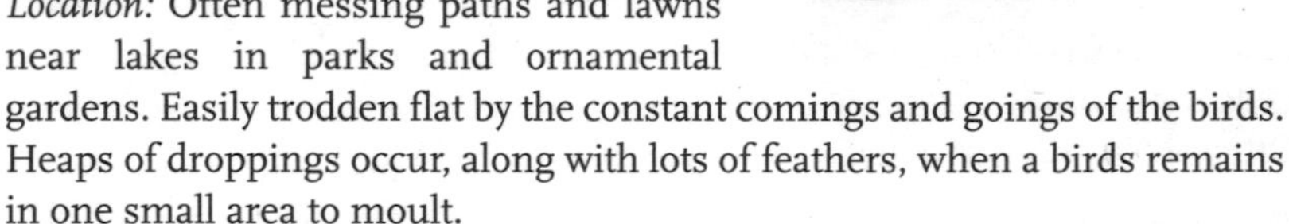

Ecological notes: Blamed for raising the nutrient levels in the water, increasing aerobic bacteria content, clouding the water and reducing the invertebrate biodiversity of waterbodies. Geese are often artificially fed bread by unwitting members of the public, who are unaware that their apparently 'helpful' activity is actually harming the environment.

Domestic chicken, ***Gallus gallus domesticus***

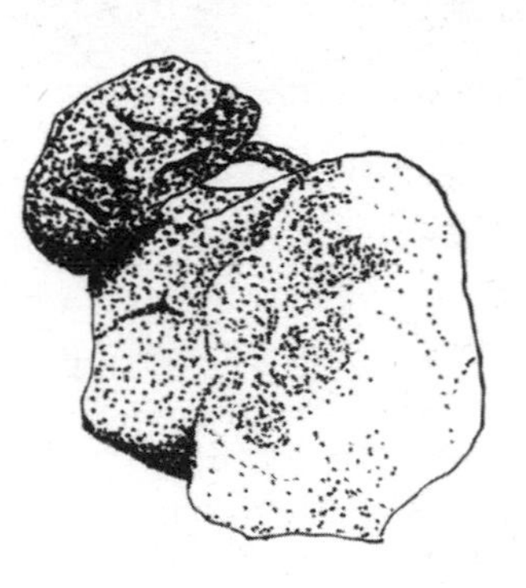

Size: 25–35 mm across, turkey droppings larger, to 50 mm across.

Description: Knob caused by deposition of convoluted cylinder of soft consistency, often with a central ball and upstanding annex. Majority of the initial extrusion coated in white uric acid, but with the tail end darker. Greenish to blackish, depending on diet.

Location: Dropped wherever across the farmyard, in coops or under the perches.

Ecological notes: A broody hen, waiting on the nest and trying to hatch her eggs, makes one or two large droppings each day, rather than many small ones. She leaves the eggs briefly and deposits in a large pat-like pile nearby before returning to her nest. Droppings with bubbles (diarrhoea), blood or worms can be used to diagnose medical problems in the hens. Despite the implied insignificance of chicken excrement, guano-style fertiliser from large battery chicken companies is now a huge offshoot of the egg and chicken meat industry.

Pheasant, ***Phasianus colchicus***

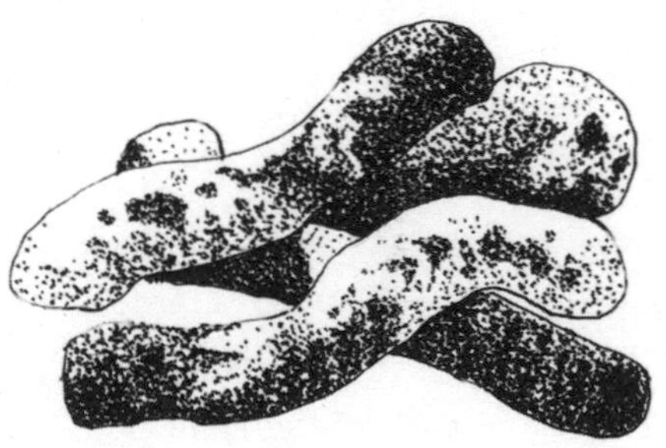

Size: 2 cm long, 4–5 mm diameter.

Description: Dark brownish grey, varying from black to greenish, tipped with white fading to grey. Spiral or curled, cylindrical shape.

Location: Woodland clearings, hedgerows, rough grassy places. Often clustered together at a roosting site.

Ecological notes: Being mainly a grain- or plant-feeding bird, the droppings are relatively firm and dry compared to typical bird splashes. Those of grouse, capercallie, partridge and peafowl are similar in form.

Heron, various genera

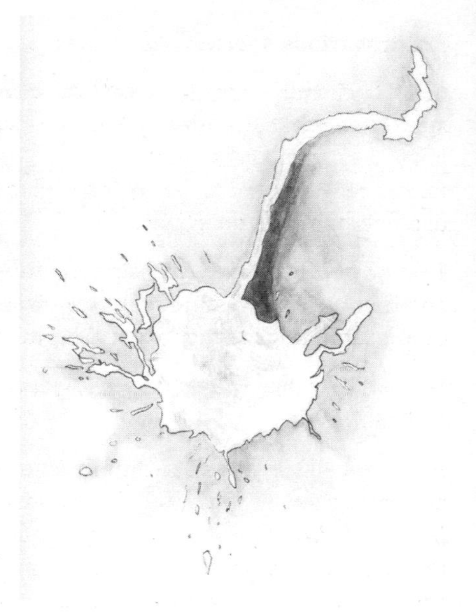

Size: 10–20 cm across.

Description: A broad splatter of black, grey and white; pungent, often strongly smelling of fish. Once described, by a victim caught in the splash, as being like having a mug of warm, fish-flavoured yogurt thrown at you.

Location: Under roosts and nesting areas (usually in trees at or near the waterside), or as it feeds on the bankside or in the shallows.

Ecological notes: Typical of predatory birds, the droppings are highly liquid splashes, since much of the feather, bone, skin and scales of their prey has already been regurgitated through the beak as a tough fibrous pellet (see owl, below). The very high nitrogen content of the faeces, typical of fish-feeding birds, tends to kill the trees in which the birds nest. This also occurs where cormorants nest.

Green woodpecker, ***Picus viridis***

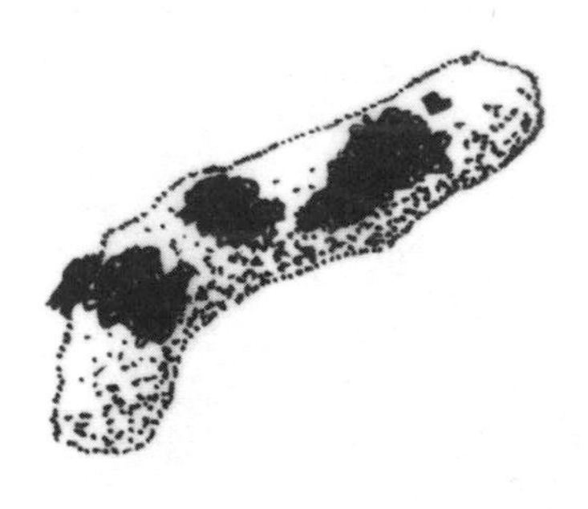

Size: 3–5 cm long, 6–8 mm diameter.

Description: Short, cylindrical package, sheathed with pale grey or white coat from uric acid, dark contents breaking out. Looks like cigarette ash and is just as brittle.

Location: Lawns and grassy areas, especially during winter when ants make up a large part of the diet.

Ecological notes: This bird feeds on insects, and specialises in eating ants (and other insects) it finds in lawns and grass root-thatch, rather than on vertical tree-trunks. The undigested exoskeleton shells of the ants can usually still be identified to species.

Snake, various species

Size: Depends on the size of the snake, 20 mm long, 4–7 mm diameter for my pet garter snake 60 cm long; up to a large deposit 30 cm long, 5–8 cm diameter for huge anaconda.

Description: Cylindrical, but variously mushy or semi-solid, twisted, spiral or amorphous blob. Mixture of dark brown, black or oily blue/green, with pale grey and white from the uric acid segment.

Location: Dropped at random, sometimes outside a burrow, or beside sunbathing site.

Ecological notes: Closely resembles bird droppings, since snakes and birds both have a single vent where bodily waste (faeces) and kidney-filtered waste (urea/uric acid) combine in joint storage cavity (cloaca) before being voided in a single movement. Some snakes withhold defecating for some time (420 days quoted for one Gaboon viper); the suggestion is that the retained weight increases the snake's fighting and subduing advantage when tackling prey.

Lizard, various species

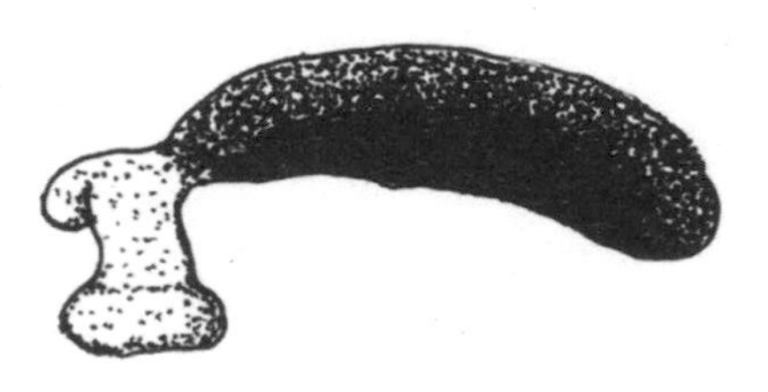

Size: 5–100 mm long, 2–20 mm diameter depending on species.

Description: Cylindrical, smooth, usually dark, especially if full of insect remains, splashed with white or with pale grey tip.

Location: Dropped more or less at random, but sometimes found on log or rock where the animal has been basking.

Ecological notes: Lizards, being closely related to birds, have but a single vent, so faeces and urine are voided together from the cloaca in a combined mass, closely resembling bird droppings. Lizard dung made headline news in 2009 when the University of Leeds threw out a bag of the stuff – only to then discover that it was a treasured collection of one of their scientists who had accumulated 35 kg of butaan lizard droppings, part of a PhD study into this rare species.

Caterpillar – frass

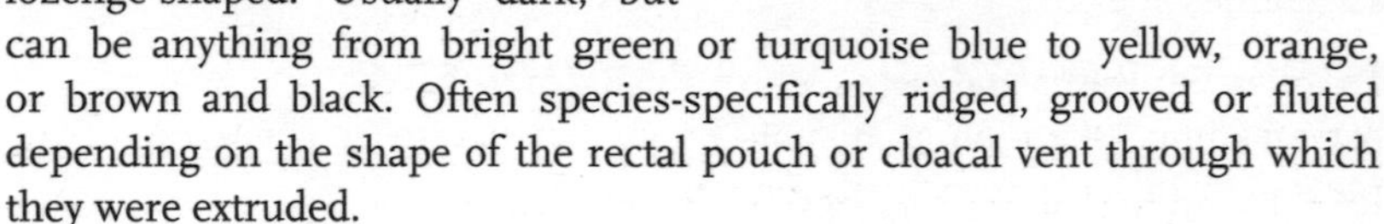

Size: 0.1–5 mm long.

Description: Short, stout, barrel- or lozenge-shaped. Usually dark, but can be anything from bright green or turquoise blue to yellow, orange, or brown and black. Often species-specifically ridged, grooved or fluted depending on the shape of the rectal pouch or cloacal vent through which they were extruded.

Location: Dropped down onto the ground, so usually invisible, but street trees with heavy caterpillar loads often have wind-blown drifts of the black pellets darkening the pale paving stones beneath. Woodlands can have constant audible rain of tiny pellets falling down and rustling on the dead leaves and undergrowth. Large frass particles are a good sign that a pot-plant or pre-packaged shop-bought culinary herbs has a caterpillar at work.

Ecological notes: Caterpillar droppings in woodland have been quoted at half-a-tonne per hectare, but this seems a rather light finger-in-the-wind estimate. In Asia, expensive caterpillar tea is produced by farm-rearing the larvae on tea plants and harvesting the frass, claiming the beverage has a smoky fermented tang.

Earthworms, *Lumbricus terrestris* and other species – casts

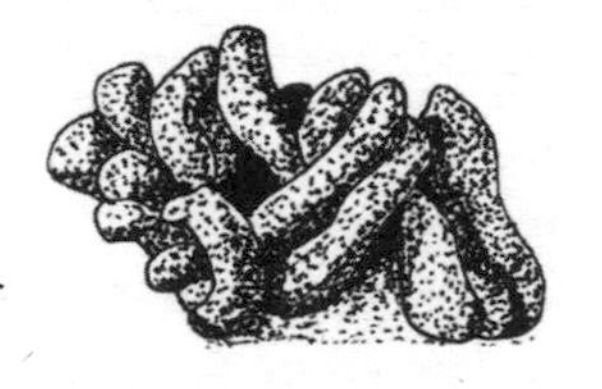

Size: Generally 35–65 mm across, 15–25 mm high.

Description: Swirl, or series of rough spirals, of finely compacted soil particles extruded into a long thin convoluted cylinder.

Location: At the entrance to the burrow, most obvious on close mown lawns, cracks between paving stones under which they are burrowing, or amongst detritus around which they are living.

Ecological notes: Large earthworms eat plant material, often dead leaves or grass stems, pulled down into their burrows, but smaller species graze minute organic growths of alga, lichen, fungal hyphae or small particles of decaying organic matter. Some soil is also ingested. When this is voided as a wormcast, it brings the non-organic matter up to the surface of the root thatch, and there is a constant cycling and recycling of top soil. Similar casts are visible at low water on marine mudflats where burrowing lugworm (*Arenicola marina*) and ragworm (many species) feed in the loose silt, although these are washed away by the returning tide.

Snail, various species

Size: 1–50 mm long, 0.1–3.0 mm diameter.

Description: Narrow squiggles, dark brown or black, sometimes streaked.

Location: On leaves and stems where the animals have been feeding. Often found in knots where they have been sheltering under plant pots, inside garden sheds, on fences, etc.

Ecological notes: Minute and unassuming to humans, snail parasitoid flies, however, use the smell of snail droppings to locate their hosts, on or near which they then lay their eggs. The lesser bulin snail, *Merdigera obscura*, coats its shell in a mixture of soil particles and its own dung, to camouflage itself.

Owl – pellets or castings (falconry term)

Size: 3–10 cm long, 15–35 mm diameter.

Description: Pale to mid grey, smooth but irregular surface, papery feel to the outside, but fibrous with matted hair and feathers inside, and containing many small bones or insect shell pieces.

Location: Often clustered together at the foot of a regular perch or roost such as a fence post, tree stump, broken wall or large pendulous tree branch.

Ecological notes: Technically pellets are not dung because they have been regurgitated back up through the mouth or beak, rather than digested and passed behind. Many birds, but especially birds of prey, cough up partly digested meal remains in the form of a compacted, sausage-shaped pellet. These contain relatively indigestible items such as large bones, fur, feathers and the occasional leg-ring from a prey item. Owl pellets are perhaps the most often found, but kestrels, hawks, falcons, buzzards, gulls, magpies, rooks and crows produce similar packets. Because they are less offensively smelly, dissecting pellets is a much easier and more socially acceptable way to ascertain a bird's diet. In Cheshire apparently called boggarts' muck, the small bones being those of fairies eaten by the aforesaid boggarts.

CHAPTER 12

DUNG INHABITANTS AND DUNG FEEDERS – A ROGUES' GALLERY

The following list is not exhaustive; instead it is a guide to the types of insects (and a few other creatures) associated with dung. The world fauna of scarabaeine dung beetles alone is something of the order of 6,000 species, clearly beyond the scope of this book. There would come a point at which this would become a repetitive catalogue. Not very readable.

Identification to species level, for all the families and genera listed below, often requires meticulous examination of specimens under the microscope and the use of complex identification keys and detailed monographs. Instead I offer here a rogues' gallery of the usual suspects. As well as feeding on or in the dung itself, associates feed on other dung-recyclers, on each other, they shelter under the dung, rest on top of the dung, or use it as bait.

I acknowledge that this list is rather Anglocentric, skewed in its content to north-west Europe, and heavily influenced by my own personal experience of dunging in the grazing meadows of southern England, but some New World and tropical species are also included to give a bit more breadth. Unless otherwise stated most species occur broadly across northern hemisphere Eurasia.

DIPTERA – FLIES

Hornet robber fly, *Asilus crabroniformis*

Size: Body length 20 mm.

Description: Large dark-brown fly, abdomen nearly black, with bright yellow tail tip; wings smoky, legs bristly yellow, head with prominent pointed snout. Hornet mimic, but harmless to humans, lacks sting.

Life history: Secretive larvae in soil layer, foodstuff uncertain, but possibly preying on other small soil invertebrates, take at least 2 years to reach maturity; adult flies predatory, attacking flies, beetles, grasshoppers or other insects, usually in midair, and killing with powerful skewer mouthparts.

Ecological notes: Adults (summer and autumn flying) sit on dry cow-pats (sometimes bare soil patches or twigs), using them as launch pads from which to attack insects flying past, including dung flies and dung beetles heading in towards the pat. Eggs are thought to be laid in the dung (cattle and rabbit recorded), leading to suggestion that larvae are specialist predators of dung beetle larvae, but no other robber flies (large world-wide family Asilidae) are dung-associated; study is still ongoing.

Snipe fly, *Rhagio scolopaceus* (and other species)

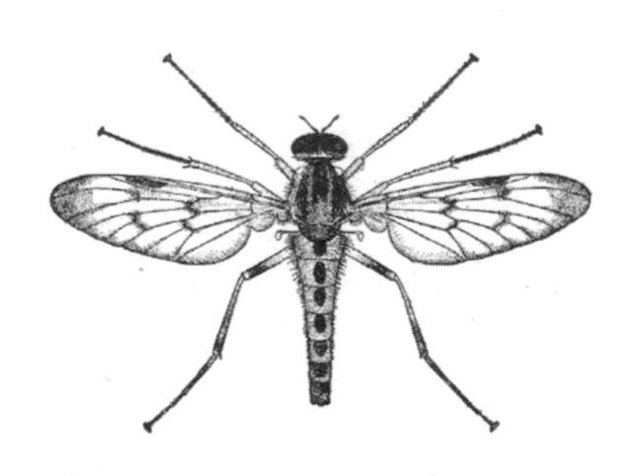

Size: Length 12 mm, wing length 10 mm.

Description: Grey fly, with long legs and conical abdomen clearly marked with yellow. Wings clear, but with brown and orange mottle pattern.

Life history: Predatory larvae, pale, worm-like, live in soil, leaf litter, wood mould, occasionally invade old dung. Adults are not known to feed.

Ecological notes: Adult flies have distinctive alert stance, resting head-down on tree trunks and other vertical surfaces. Intuitively they look as if they are poised for hunting forays, but these are usually males, and the only hunting they do is after the females which mostly rest on the ground.

Yellow dung fly, *Scathophaga stercoraria*

Size: Body length 5–6 mm.

Description: Furry or bristly fly, dull greenish grey in female, but densely bright golden furred in male.

Life history: Eggs laid in fresh dung (cow-pats especially preferred); pale whitish-grey, slim, conical maggots feed in the dung for 10–20 days then pupate in the soil; adults emerge 10–80 days later depending on temperature. Adults predatory on flies and other small insects.

Ecological notes: One of the most obvious and important dung insects. Fresh dung, minutes on the ground, soon attracts males which flit about actively and aggressively waiting for females. Mating takes 20–50 minutes, but male remains on female's back 'guarding' her for similar (or longer) time, to prevent other males diluting his sperm through multiple copulations. Female leaves after egg-laying, so active pats often male-dominated, leading to squabbles over incoming receptive females. Wetter or looser dung types preferred: cow or horse in meadows, but cat, dog or fox in urban parks and gardens. Several very similar species.

Black 'ant' flies, *Sepsis* and other genera

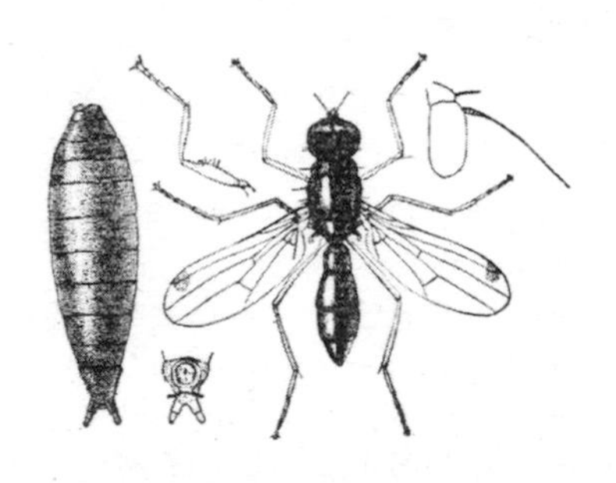

Size: Body length 2–6 mm.

Description: Tiny, black, shining flies, with relatively long legs, slim waist and round head, giving an ant-like appearance. Clear, narrow wings tipped with small but distinct black dot near apex.

Life history: Breed in soil, dung, decaying vegetation and other rotten organic matter.

Ecological notes: Large 'swarms' of the flies (many thousands) can emerge and crawl/fly about in the grass where the cow-pat or other dropping has long gone. As they crawl over the vegetation, they flick their spotted wings about in territorial or courtship display.

Moth flies, family Psychodidae

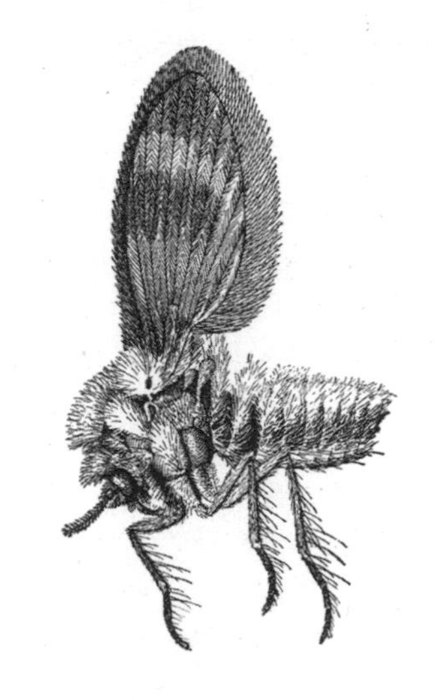

Size: Length to 2.5 mm.

Description: Tiny, slightly fluffy, flies with short, broad wings, also covered or patterned with fluffy scales. Wings held flat over back in delta shape, hence moth name. Various species coloured black, through brownish mottles to pure white.

Life history: Tiny pale maggots feed in wet detritus, soil, rotting fungus, leaf litter, manure, compost and dung.

Ecological notes: Often found swarming in clouds close over and around dung. Small size means they can crawl between the pellets or folds of the dropping. They are a major inhabitant of trickle beds in sewage treatment works. Some species breed in drains, including in the algal film growth in domestic sink overflows in kitchen or bathroom, but not implicated in disease spread.

Daddy longlegs, or craneflies, *Tipula* species

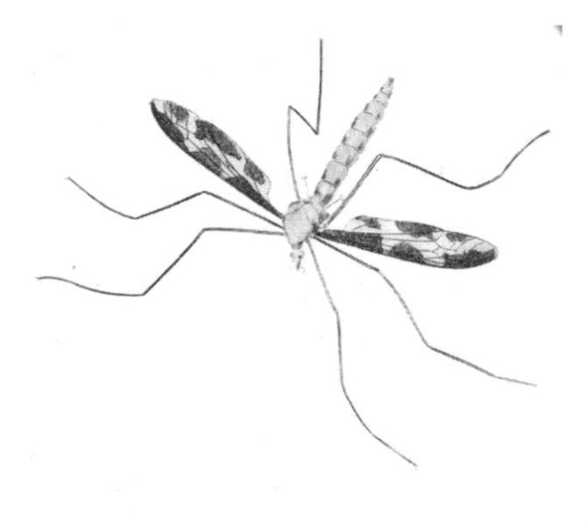

Size: Wing length to 30 mm, body length to 35 mm, leg span to 55 mm.

Description: Narrow-bodied, narrow-winged, long-legged flies, various shades of grey, sometimes with wings smoky or marked with dark mottled patterns. Adults fly with ungainly bobbing flight, often half crawling, through the long grass. Some species attracted to porch lights.

Life history: Stout cylindrical larvae ('leatherjackets') are grey wrinkled and short worm-like, feed in soil, leaf litter, root thatch, on living and dead plant material.

Ecological notes: Not really dung specialists, but larvae often found under old dung where soil horizon is blurred.

Biting midges, *Forcipomyia* species

Size: Wing length to 4 mm.

Description: Small to minute grey or blackish flies. Bodies slightly hunched, wings clouded with dusting of microscopic dark scales. Males have distinctive plumed antennae.

Life history: Pale grey bristly larvae live in soil, dung and leaf litter, eating decaying organic matter, fungi or algae.

Ecological notes: Although part of 'biting midge' family, are inoffensive to humans or stock animals, probably biting and sucking body juices from other, larger, insects. A *Forcipomyia* midge achieved the rare distinction of having the highest recorded wing beat of any insect, at over 2,200 beats per second. However, this was only in the laboratory, and after the researcher had heated up the unfortunate fly, and cut most of its wing membrane off.

Biting midges or 'punkies', *Culicoides* species

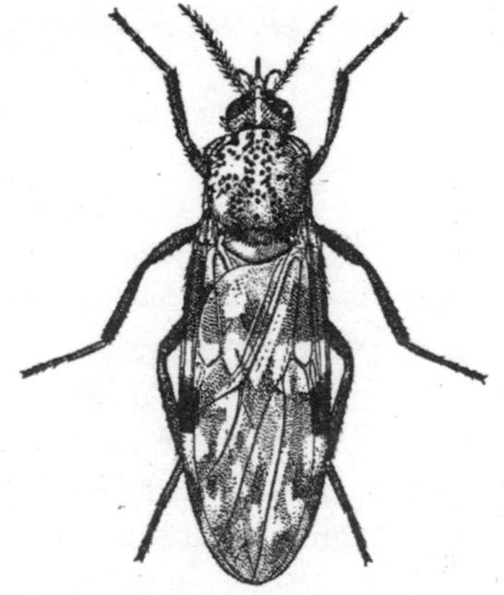

Size: Wing length to 6 mm.

Description: Small to minute grey, brown, black or yellowish flies. Body stout and hunched, wings often patterned with dark mottles. Male antennae thickly plumose.

Life history: Tiny pale worm-like larvae feed in dung, or waterlogged soil, eating fungi, algae or decaying organic material.

Ecological notes: Adult flies bite and suck blood from humans and farm animals. These are the midges that make upland areas intolerable for much of the year. In North America, they are blamed for huge financial losses to the tourist industry because their flight season seriously curtails outdoor activities. *C. imicola* is main vector implicated in spread of bluetongue virus among sheep in Europe.

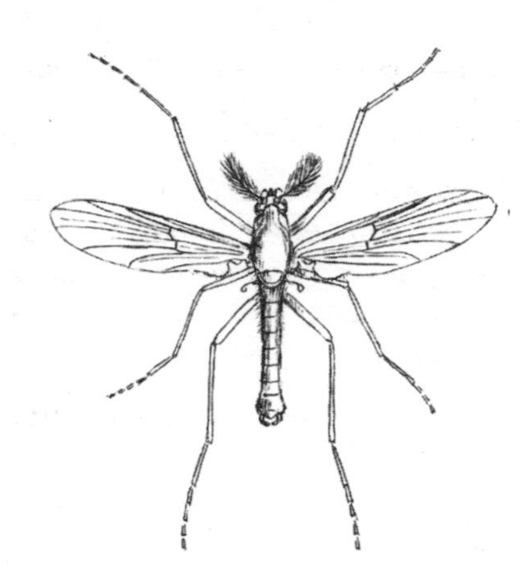

Non-biting midges, family Chironomidae

Size: Length to 10 mm, wing length 8 mm.

Description: Narrow-winged, narrow-bodied, long-legged flies, various shades of grey, brown or black. Males with broad feathery antennae. Closely resemble mosquitoes, to which they are related, but lack any biting mouthparts.

Life history: Most species are aquatic, larvae living in the mud at the bottom and sides of ponds, ditches and slow streams, feeding on decaying organic matter. Bobbing midge clouds over water are often members of this large family.

Ecological notes: Some species renowned for their ability to live in low-oxygen muds; these larvae contain the strongly red-coloured oxygen-storing chemical haemoglobin (the same as found in mammalian blood), giving them a bright scarlet colour and earning them the name bloodworms. Not surprisingly a few species have been reared from dung.

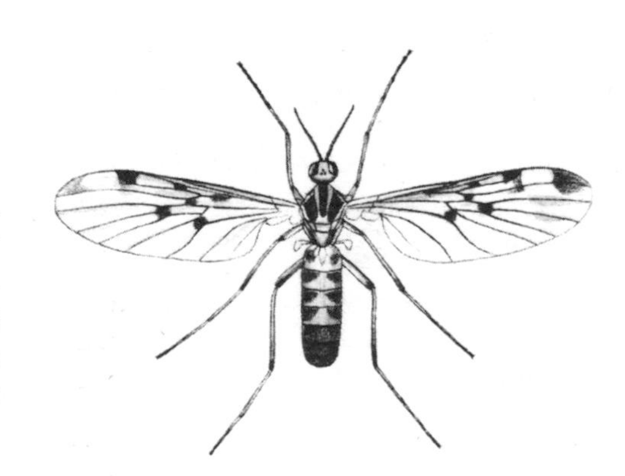

Window gnat, *Sylvicola punctata*

Size: Wing length to 6 mm.

Description: Small, greyish brown, mosquito-like fly with tiny spherical head. Wings marked with dark cloud of mottles, sometimes approximating to three dark blotches.

Life history: Slim, worm-like larvae live in soil, dung and leaf litter, eating fungi, algae and decaying organic matter. Adult flies mostly nocturnal.

Ecological notes: Gets its common name from its habit of appearing on the inside of domestic windows after emerging from pupae in the soil or compost brought in with potted house-plants, but does not visit food and is not implicated in any disease spread.

Fungus midges, family Sciaridae

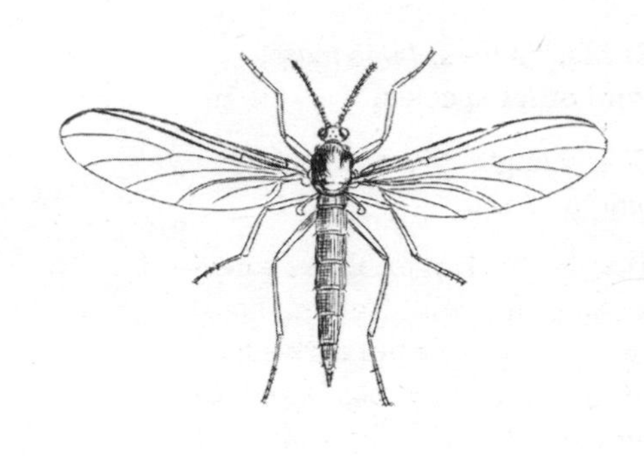

Size: Wing length to 3.5 mm.

Description: Small, dark, delicate flies with long legs, long antennae, long wings and narrow bodies. Wings often darkened. Fly slowly, but scuttle quickly. Tibiae of all legs with long distinct spurs at ends.

Life history: Slow-moving larvae feed on fungi, in soil, in decaying organic matter.

Ecological notes: Often found breeding in old dung when fungal decay has set in. Sometime pest in mushroom farms where they can breed in the rotted manure-based compost, but also potentially attack the hyphae of the crop, or invade any overripe mushrooms on the turn.

Dung midges, family Scatopsidae

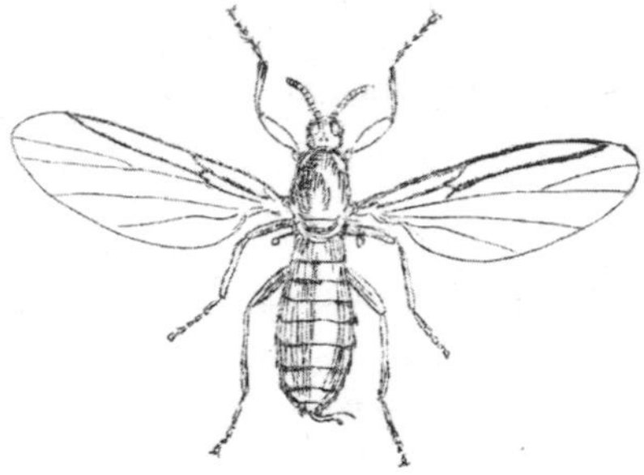

Size: Length to 3 mm, wing length 2.5 mm.

Description: Minute dark or black flies with seemingly over-long wings or extra short bodies. Abdomen short oval, legs short, wings whitish with veins unpigmented, antennae short and stout.

Life history: Slow-moving bristly larvae breed in soil, leaf litter and dung. Adults often congregate in large numbers on flowers.

Ecological notes: A poorly studied group of minute flies, little is known about their habits or life histories as demonstrated by their other common name 'minute black scavenger flies'. In marshy areas of grazing meadows they sometimes form clouds of many thousands skittering about in the long grass, or over scrub.

St Mark's flies, *Bibio marci* (and other species)

Size: Length to 15 mm, wing length 12 mm.

Description: Large, black, sooty-looking fly, with robust body. Broad wings pale but dark-edged in male, heavily blackened in female. Flies heavily, with long black legs hanging down.

Life history: Pale larva, live in the soil, feeding on decaying (possibly also living) plant matter, so regularly invade old dung where humus horizon is blurred.

Ecological notes: Named for appearance, often in large clouds, around St Mark's day, 25 April, although this is very approximate. Many similar species, some with bright red front legs. Smaller soil-inhabiting bibionids often called fever flies, for no very sound reason, can occur in huge numbers; larval densities of 37,000 per square metre are recorded.

Broad centurion, *Chloromyia formosa*

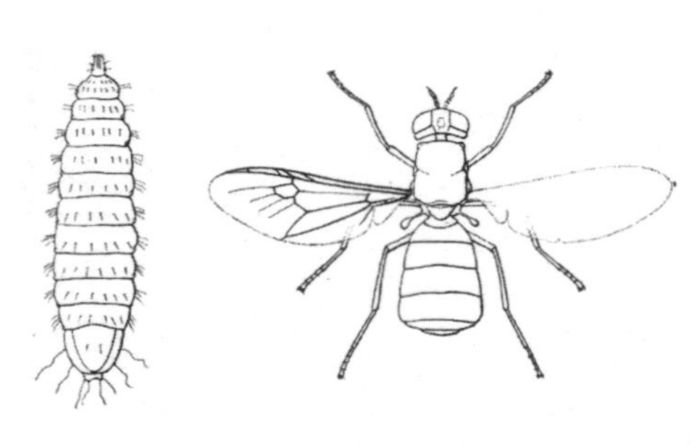

Size: Length 9 mm, wing length 7 mm.

Description: Broad, flat, parallel-sided, metallic green to violet soldier-fly, abdomen of males golden or bronze, wings broad and delicate, veins hardly visible, legs and antennae short.

Life history: Short, stout grey larva lives in dung, damp soil and wet leaf litter, feeding on decaying organic matter.

Ecological notes: Really a denizen of damp soil, but regularly occurring in the blurred boundary between humus and old dung. Common in fields, woods and gardens. Gets soldier name because related species sometimes brightly coloured like military dress uniforms.

Black-horned gem,
Microchrysa polita

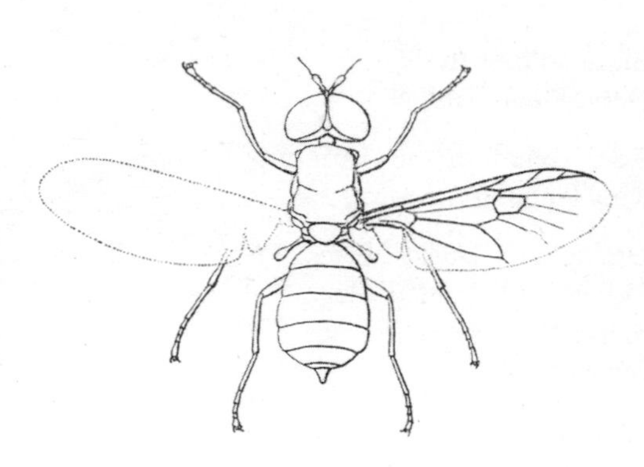

Size: Length 5 mm, wing length 4 mm.

Description: Tiny metallic green, or bluish, soldier fly, abdomen short, squat, parallel sided, almost hexagonal. Legs and antennae black (yellow in other species). Wings delicate, clear, veins hardly visible.

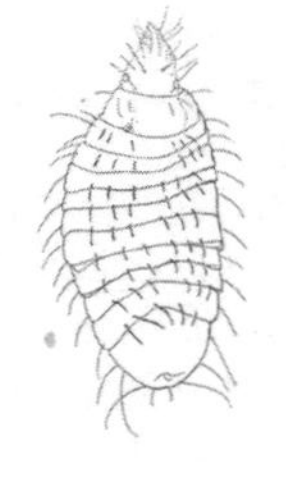

Life history: Tiny grey larva short and stout, breeds in and under fresh dung, in manure heaps and compost bins.

Ecological notes: Needs moist, putrescent organic matter. Many other species of soldier fly are aquatic, breeding in muddy pond edges, ditches and slow streams.

Twin-spot centurion,
Sargus bipunctatus

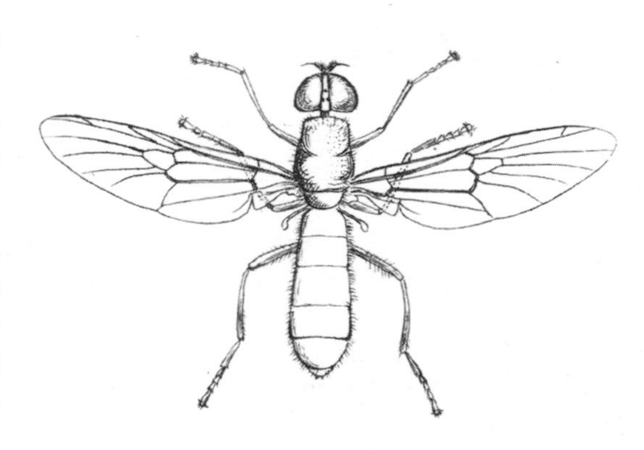

Size: Length to 10 mm, wing length to 9 mm.

Description: Narrow, parallel-sided body, metallic green thorax but abdomen golden bronze in male, dark metallic blue with pinkish-orange girdle in female. Legs yellow.

Life history: Several similar species breed in wet meadows, marshy woods, in the damp soil, larvae feed on decaying organic matter.

Ecological notes: The most closely dung-associated of this large genus in Europe; often the only species found in towns and cities where it regularly breeds in cat, dog and fox dung.

Black soldier fly,
Hermetia illucens

Size: Length to 16 mm, wing length to 14 mm.

Description: Dusky black fly, wings clouded, legs marked with white or yellow.

Life history: Native to south-eastern USA, breeding in compost and manure, animal dung and carrion. Larvae ('phoenix worms') are dung/decay feeders.

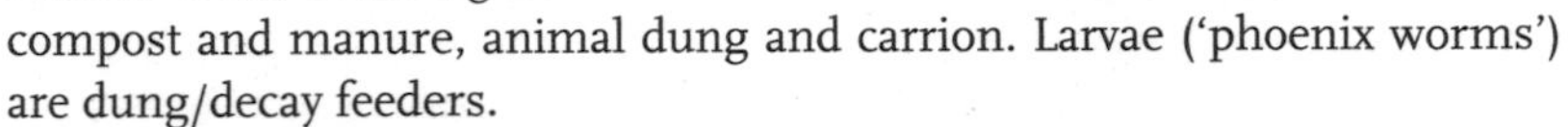

Ecological notes: Introduced throughout much of the northern hemisphere, either because maggots are supplied by pet shops as lizard or fish food, or as part of manure control in farms. The larvae are amenable to captive rearing on an industrial scale to process chicken or pig manure. The larvae are then edible to the livestock – a particularly neat recycling of nutrients. Also potentially edible by humans, but in processed ground meal form.

Long-footed flies, *Dolichopus* species (and many similar genera)

Size: Length to 6 mm, wing length 5 mm.

Description: Small greyish, or often metallic greenish, flies with slim legs on which the short body seems to perch, head up, tail down. Active and fast, scuttling and skipping about.

Life history: Predatory as adults, hunting and killing other small flies. Larvae in soil, mud, rotten wood, standing water; most are predatory eating small invertebrates.

Ecological notes: Not really dung-feeders, but will visit dung to prey on the other small flies it attracts.

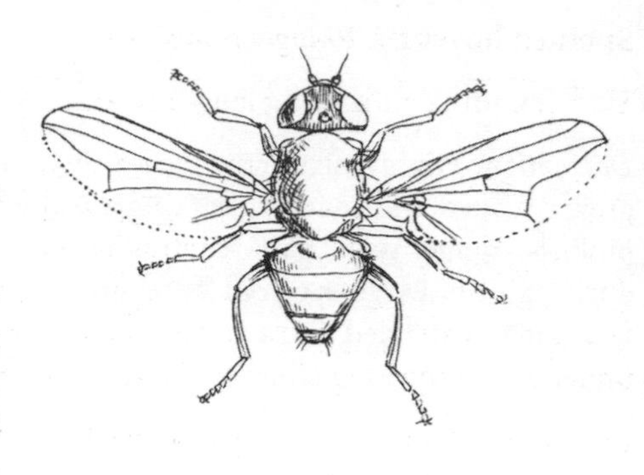

Metallic scavenger fly, *Physiphora alceae* (formerly *Chrysomyza demanata*)

Size: Length 5 mm, wingspan 9 mm.

Description: Small, squat, short-bodied black fly with distinct metallic green reflections. Head red, eyes prettily striped red and orange. Wings pale, clear. Female abdomen pointed with telescopic egg-laying tube.

Life history: Breeds in manure heaps, compost and other mainly decaying plant matter, but also sometimes recorded on dung.

Ecological notes: An elaborate courtship dance has been reported, involving male leg-waving and wing-flicking.

Lesser dung flies, family Sphaeroceridae (formerly Borboridae)

Size: Length to 4.5 mm, wing length 4 mm.

Description: Tiny to very small black or dark grey flies, rather hunched, scuttle readily on their stout legs. Very many extremely similar species.

Life history: Larvae of some species are specialist dung-feeders.

Ecological notes: Adult flies will enter the dung to lay their eggs, using tunnels dug by other dung-users, mainly dung beetles. Their relatively short wings and preference for running, rather than flying, may aid them in their exploration of the dung tunnels and crevices.

Snouted hover fly, *Rhingia campestris*

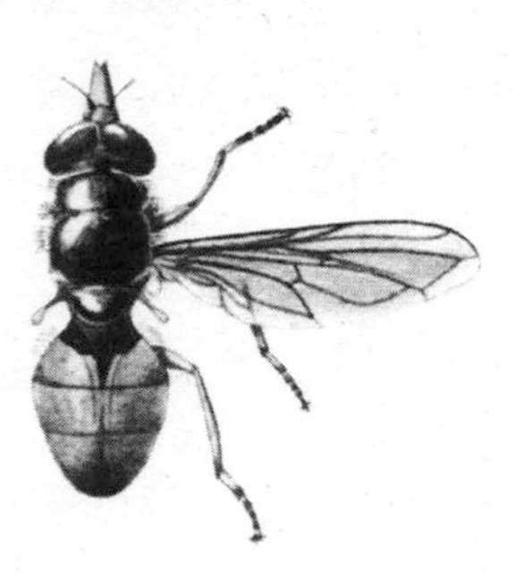

Size: Length 12 mm, wing length 10 mm.

Description: Highly distinctive hover fly, thorax grey, streaked with darker markings, abdomen pinkish-orange with dark central stripe and dusky tail tip. Large eyes red, head greyish, but face hugely extended into a narrow orange 'beak' under which the long elbowed tongue is stored.

Life history: Stout, flattish larvae feed in dung, cow especially, and also sometimes in manure or compost heaps, and probably in other decaying organic matter.

Ecological notes: Adults rarely (if ever) found in the immediate vicinity of dung, and seemingly visit only to lay eggs, then depart. They are mostly found resting on herbage or visiting flowers and can use their long tongue to reach deep into plants with long corolla, which other flower visitors cannot reach. For a time this insect was called the Heineken fly, echoing the catchphrase of a long-running and highly successful 1970s advertising campaign, where the beer was able to 'refresh the parts other beers cannot reach'.

Thick-legged hover fly, *Syritta pipiens*

Size: Length 6 mm, wing length to 6 mm.

Description: Slim, narrow hover fly; body dark, but abdomen clearly patterned with four yellow triangles in male, or two yellow triangles and four white tick marks in female. Face white, head dominated by large dark eyes. Legs dark brown, but marked with orange. Femur of hind leg grossly enlarged into bulbous swollen segment, with ridge of small sharp spines running beneath it.

Life history: Larvae in wet organic matter, including compost and manure heaps, silage and dung.

Ecological notes: Adults never seen at dung, but ubiquitous visiting flowers, where large numbers can congregate, making it one of the commonest of insects. Superb hoverers, males engage in head-to-head hovering jousts, making sudden darting movements (to other insects too), retreating and dodging, before eventually one retires and the other continues patrolling. Reason for thickened legs unknown.

Drone fly, *Eristalis tenax*

Size: Length 15 mm, wing length 13 mm.

Description: Glossy black or dark brown hover fly, stout body marked with orange-brown triangles or dashes on sides of abdomen. Eyes very large, dominating head. Closely resembles a honeybee (especially the large-eyed males – drones), for which it is frequently mistaken, especially by journalists and subeditors.

Life history: Semi-aquatic larva lives in watery detritus, has long rat-tail telescoping breathing tube to reach surface for air. Adult flies active all year, visiting flowers.

Ecological notes: Not strictly a feeder in dung, but usually in ditches where nutrient-rich run-off has entered from sewers, manure heaps, farmyards, slurry pits. This is the *bugonia* or oxen-born bee of the ancients who thought honeybees could spontaneously generate from the carcass of a cow, not realising that *Eristalis* was quite at home breeding in the putrescent semi-liquid decay of the corpse. The wasp-mimicking black and yellow *Myathropa florea* occurs in similar places.

Face fly, *Musca autumnalis*

Size: Length 7–8 mm, wingspan 13–18 mm.

Description: Superficially similar to house fly in shape, though slightly larger and sturdier. Grey, thorax dark-streaked, abdomen tessellated with chequer pattern of light and dark greys in female, male abdomen orange with dark central streak and ticks down back. Wings clear, but stained orange near bases. Eyes dark red.

Life history: Yellowish-white maggots feed in manure, slurry and dung. Adults visit flowers, but also torment stock by trying to sip liquid around the eyes and mouth.

Ecological notes: Will enter houses in autumn (hence scientific name), but only to hibernate, so not implicated in disease spread onto human foods. Does, however, spread eyeworm (nematode) and pinkeye (infectious conjunctivitis) in cattle. Walkers and ramblers through cattle pasture can be pestered by clouds of them flying around the head.

Bush fly, *Musca vetustissima*

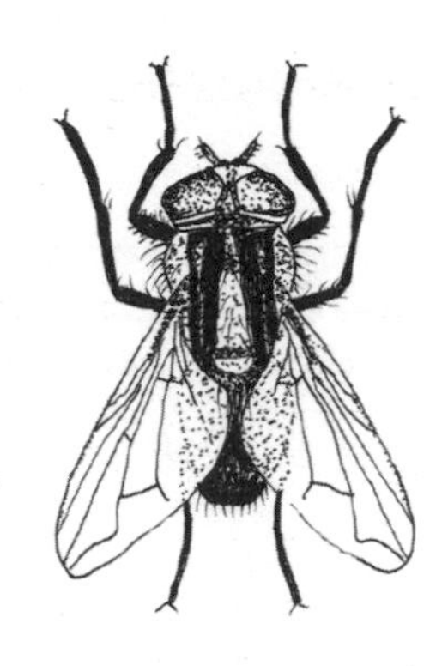

Size: Length 6–7 mm, wingspan 13–15 mm.

Description: superficially very similar to house fly. Grey mottled fly, thorax with two broad darker streaks, abdomen greyish.

Life history: Australian. Breeds in moist decaying organic matter, and although not overly fond of native marsupial droppings because they were too dry, it thrived in the semi-liquid droppings of cows when these were introduced by European settlers.

Ecological notes: Particularly a problem in summer (October–March) when the air is thick with clouds of them. There are regular tales of travellers in the bush being unable to speak because immediately they opened their mouths in the swirling maelstrom, flies flew in. This gave rise to the stereotypical Aussie bush-hat with corks dangling on strings, and the 'Australian salute' flick of the hand across the face. It was partly to control bush fly numbers that African and Eurasian dung beetles were introduced into Australia in the mid to late 20th century.

House fly, *Musca domestica*

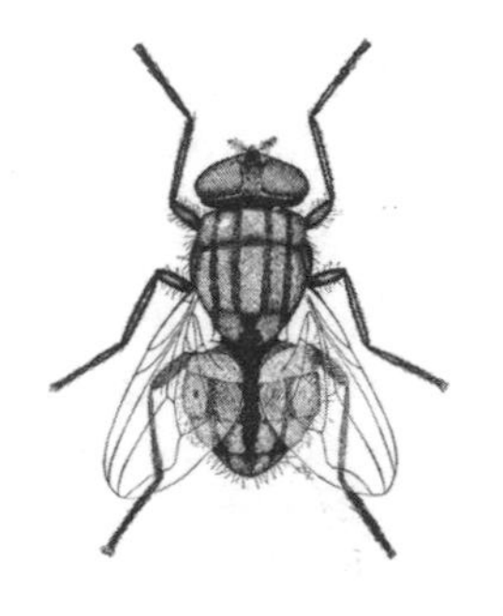

Size: Length 6–7 mm, wingspan 13–15 mm.

Description: Short and stout fly, grey, thorax with four darker streaks, abdomen with sides pale translucent yellow, legs dark grey to black.

Life history: Pale grey maggots feed in semi-liquid decaying organic matter – compost, manure, dung, carrion. Adult flies feed in the same breeding material, but then come into human houses and visit food left out. They eject enzyme juices, including contaminants from previous faecal encounters, from internal glands, then suck up digesting food.

Ecological notes: Long blamed for disease spread, directly by the method of eating by regurgitating crop contents from previous meal, but also by traipsing filth over human food. Reviled as 'the insect menace', houseflies were found to harbour more than 100 types of bacteria, virus and protozoa, and in one report over 6 million bacteria on a single fly. Advances in sanitation and food storage have improved the situation. Now much less common in towns and cities, still abundant in rural areas and really might be called the farmhouse fly.

Lesser house fly, *Fannia canicularis*

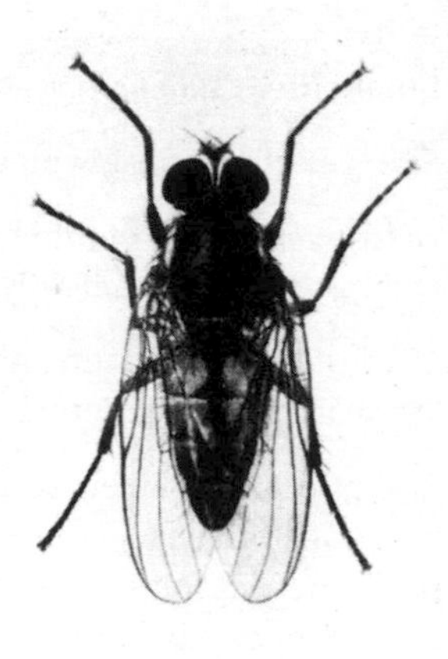

Size: Length 3–4 mm, wingspan 8–9 mm.

Description: Mottled grey fly with vaguely pale-streaked thorax, and yellow side markings on abdomen. Legs dark, wings clear.

Life history: Spiny wrinkled larvae live in almost any decaying organic matter, including compost, fungi, carrion and leaf litter in the soil. Adults come indoors and fly mad zigzags under pendent light fittings, but are not attracted to food and are not implicated in disease spread.

Ecological notes: Although this species is most often associated with old dung, where it may be feeding on fungal hyphae, the closely related *F. scalaris* has earned the name latrine fly for its very close association with human and other fresh dung. There are very many similar species with varying degrees of association with dung and manure.

Greenbottles, *Lucilia* species

Size: Length to 10 mm, wingspan 15–20 mm.

Description: Broad, stout, shining metallic green flies, sometimes with red, copper or gold tints, or appearing slightly bluish. Short legs dark. Large eyes dark red. Wings clear. Many similar species. The larger 'dead dog fly', *Cynomya mortuorum*, to 15 mm long, darker, bluer, bristlier.

Life history: Fat, pale-grey maggots feed in carrion or other decaying animal matter such as found in food leftovers dumped into compost heaps of domestic rubbish bins, or in commercial rubbish tips.

Ecological notes: Although not known to breed in dung, these attractive flies visit fresh dung to feed and can arrive within minutes of deposition. Large numbers can occur, creating a loud buzz when they are disturbed. Are attracted to meat-based human foods (fresh or cooked) so potential for disease transmission on the barbecue, picnic or open-air market stall, although not regular household visitor indoors.

Bluebottle, *Calliphora vomitora*

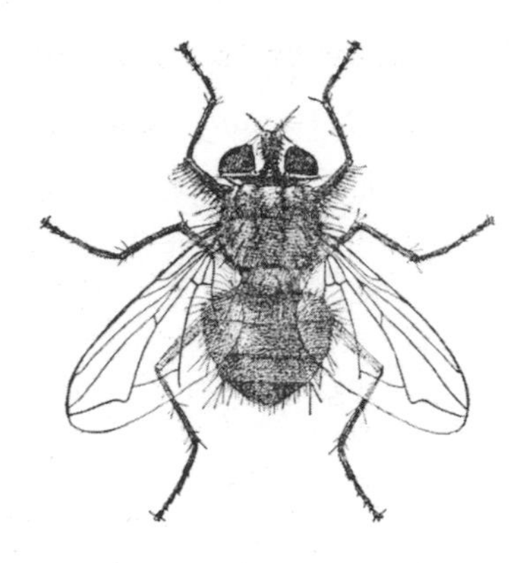

Size: Length 12 mm, wingspan, 20 mm.

Description: Large stout shining blue fly, verging on metallic, bristly.

Life history: Like its greenbottle cousins, larvae in decaying animal material.

Ecological notes: Likewise does not breed in dung, but visits fresh droppings to feed, often in loud buzzing hordes. Will come indoors, buzzing lazily about, and often called blow fly for its habit (before fridges invented) of laying eggs on stored meat, then said to be 'fly-blown'. The blowing of its name may be derived from bloated ('blowted') carrion and decaying meat. External feeding by spitting out digestive juices onto food, then sucking this back up, echoed in its scientific name; visiting human food after dung-feeding thus not a pleasant thought.

Horn fly, *Haematobia irritans*

Size: Length 4 mm, wingspan 7 mm.

Description: Small grey compact fly reminiscent of house fly. Dark grey, with vague pattern of lighter mottles. Wings clear.

Life history: Eggs laid in very fresh dung. Requirement for blood-heat freshness of the dung has been invoked; sometimes flies said to lay eggs before defecation has finished. Pale larvae hatch almost immediately, feed-up quickly and take about 7 days to mature enough to pupate. Continually brooded until late autumn in temperate areas, when some pupae remain dormant until adult emergence in spring.

Ecological notes: Adult flies suck animal (but not human) blood and can reach nuisance proportions around stock animals, many thousands per cow, causing distress and lowered milk yields. The similar stable fly, *Stomoxys calcitrans*, will (painfully) bite humans; it breeds in stable litter and manure in temperate zones (overwintering as slow-developing larva), but directly in dung in open grassland in warmer regions.

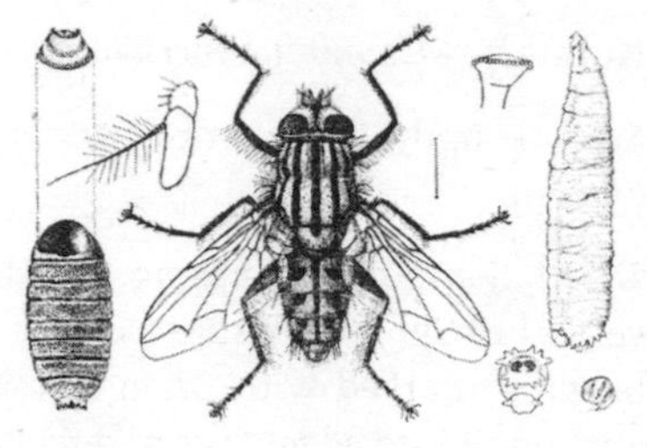

Flesh flies, *Sarcophaga* species

Size: Length to 20 mm, wingspan to 22 mm.

Description: Large, stout, mottled grey and black flies, chequered pattern shifting with the light. Eyes bright red

Life history: Mostly carrion breeders (even small items such as dead snails and insects), requiring decaying animal matter for the maggots to feed, but some dung-breeding species are known throughout the world.

Ecological notes: Even if not breeding in the dung the adult flies often occur as casuals, resting on it, and presumably attracted by the same aromatic decay chemicals.

Horse bot fly, *Gasterophilus intestinalis*

Size: Adult fly length 18 mm, wing length 15 mm; larva to 20 mm.

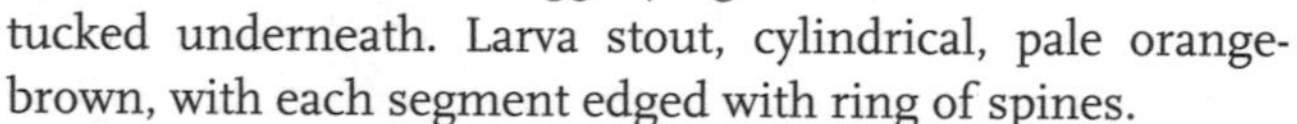

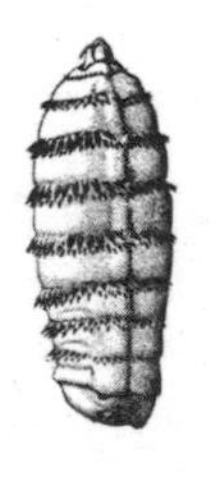

Description: Adult flies are all-over orange-brown and grey-brown, furry, sometimes said to resemble bees. Wings clear with dark blotch in middle. Female fly with narrow tail, egg-laying tube, tucked underneath. Larva stout, cylindrical, pale orange-brown, with each segment edged with ring of spines.

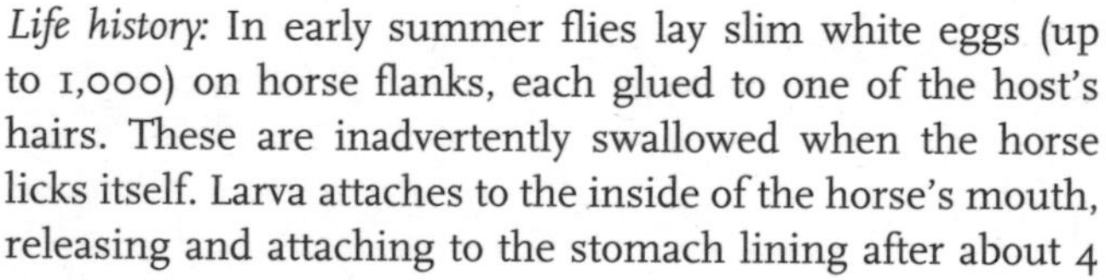

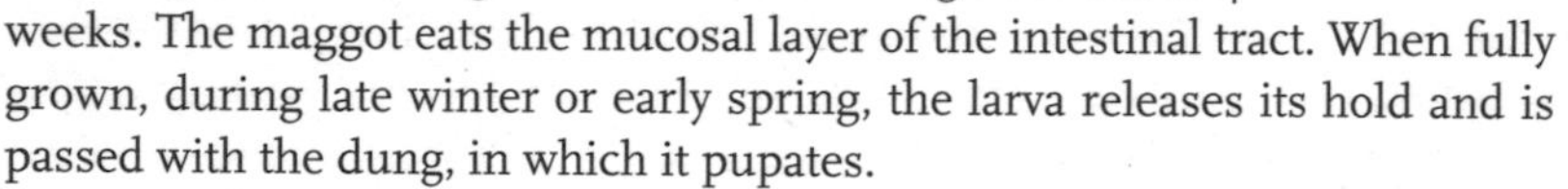

Life history: In early summer flies lay slim white eggs (up to 1,000) on horse flanks, each glued to one of the host's hairs. These are inadvertently swallowed when the horse licks itself. Larva attaches to the inside of the horse's mouth, releasing and attaching to the stomach lining after about 4 weeks. The maggot eats the mucosal layer of the intestinal tract. When fully grown, during late winter or early spring, the larva releases its hold and is passed with the dung, in which it pupates.

Ecological notes: Not a dung feeder, but larvae (also pupae) are found in the dung. Adult flies have no functional mouthparts and do not (cannot) feed. Now a very rare insect in many areas where the valuable livestock are treated with systemic chemical drugs to rid them of bots, worms, warbles and other disease-status organisms.

Noon fly, *Mesembrina meridiana*

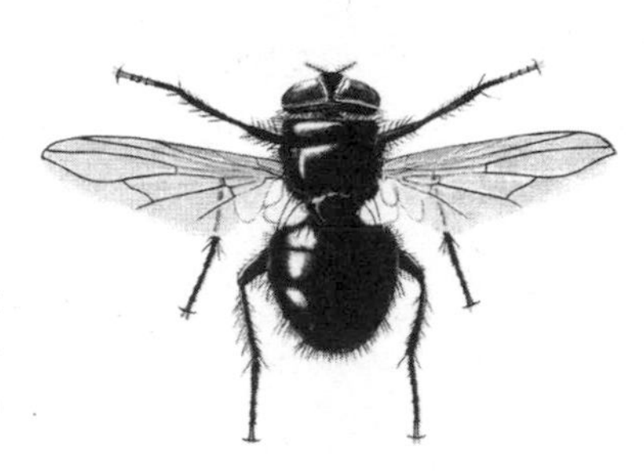

Size: Length to 12 mm, wingspan 23 mm.

Description: Large, shining, jet-black, very bristly fly. Wings clear, but brightly marked with orange at base and along fore-edge. Face and feet also bright orange.

Life history: Breeds in dung (mainly cow). Female lays only about five eggs in its lifetime, singly, often a day or two apart. Egg hatches quickly (1 hour), but if suitable pat is not found it hatches in the fly's egg-laying tube and live larva is laid instead. Deep yellow maggot (largest fly larva in north European dung pats) has large jaws and will eat other fly grubs if they are available.

Ecological notes: A spectacular and handsome insect, often found sunning itself on tree trunks, leaves or fence posts, at the height of the day, hence common name. Also visits flowers. Given the chance the larva is fiercely predatory, but can survive on a dung-only diet.

Mottled dung fly, *Polietes lardarius*

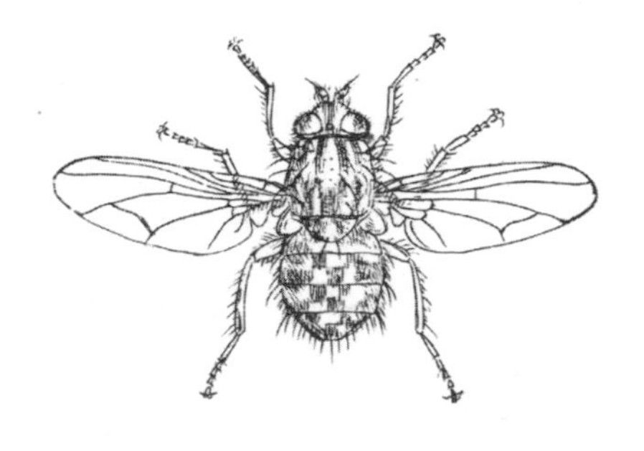

Size: Length 10 mm, wingspan 18 mm.

Description: Medium-sized, stout, broad, grey fly, prettily marked with dark stripes down thorax and black bars across abdomen, creating a harlequin-like pattern that seems to shift as the light angle changes. Legs dark grey. Large eyes dark red.

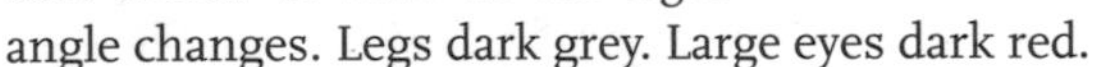

Life history: Pale larvae live in dung, and are probably predatory on other fly maggots, at least when they are near fully grown. Adults rest of leaves and tree trunks as well as visiting dung.

Ecological notes: Several similar species. *P. hirticrura* ferociously predatory and recorded taking on much larger predatory larvae of *Mesembrina* (see above). These two are cow dung specialists, but other species apparently live only in horse dung.

COLEOPTERA – BEETLES

Ground beetles, family Carabidae

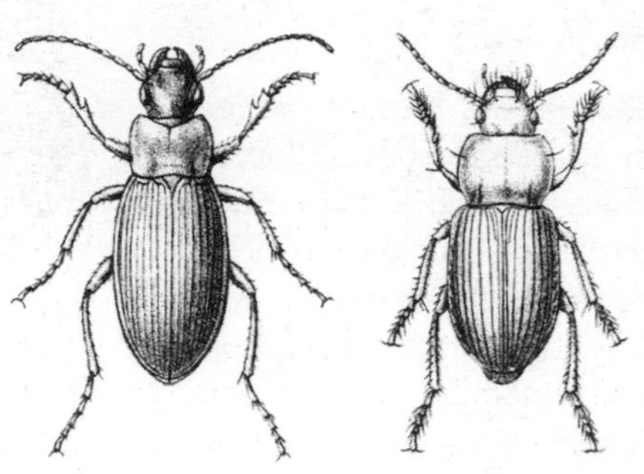

Size: 3–30 mm.

Description: Very large group of diverse beetles, generally long oval, long legs, long antennae, flattened or cylindrical bodies, shining black or dark metallic.

Life history: Predatory as dark armoured larvae, and adults. Very active, fast running. Sometimes found under dung, and would not doubt take advantage of dung insect prey if they found it, but not really part of the dung fauna.

Ecological notes: Denizens of the grass thatch and have specially strengthened hind legs to wedge-push their way through the tight stems and roots, in pursuit of prey and to escape enemies. Often found under logs and stones, and treat dung as similar shelter.

Swimming dung beetle, *Sphaeridium scarabaeoides*

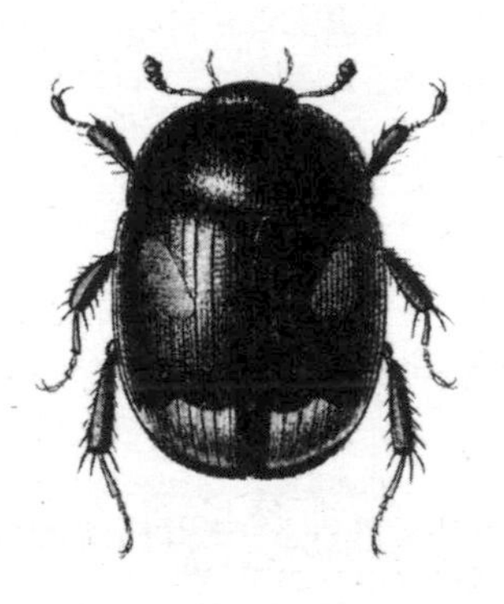

Size: Body length 5.5–7.5 mm.

Description: Shining, round, slightly domed beetle, black with reddish patches near shoulders and across tips of wing cases; legs broad and spined. Several very similar species.

Life history: Wrinkled, grey/brown, maggot-like larvae are predatory, attacking and eating other small dung- and soil-dwelling insects and their larvae.

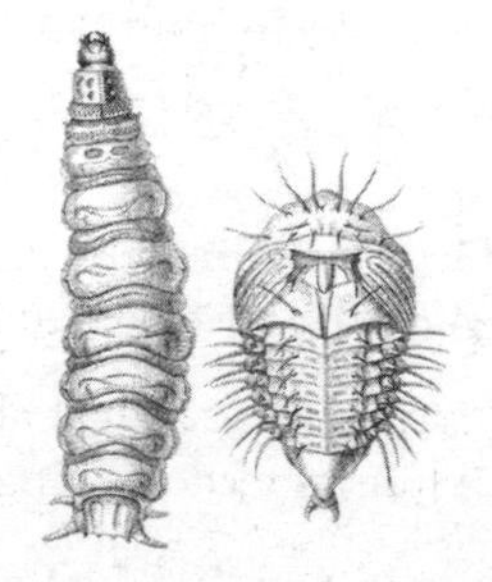

Ecological notes: Flattened, rounded and smooth outline and broad paddle-like legs allow it to 'swim' through fresh semi-liquid cow dung. Flying new arrivals to the runny pat land and break straight through the thin skin-like rind forming in the sunshine. Related beetles in this family (Hydrophilidae) live in water, or mud.

Scavenger dung beetles, *Cercyon* and other genera

Size: 1.2–4.5 mm.

Description: Minute to very small, highly convex, domed, shiny beetles. Usually black, dark brown or dirty red, but sometimes marked with brighter red or yellow. Legs and antennae short. Several similar genera include *Cryptopleurum*, which is slightly hairy under the microscope, and *Megasternum* which has very broad, deeply notched front legs.

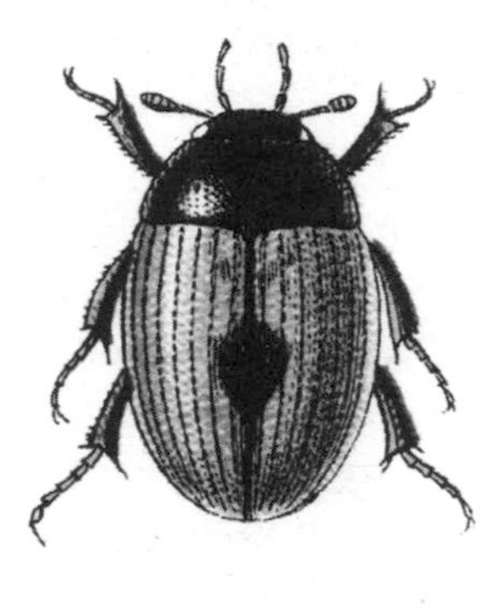

Life history: Various species breed in decaying organic matter of all types, including compost, carrion, rotting fungi, pond edges and seaweed strandlines; a few are more or less confined to dung and manure.

Ecological notes: Large numbers of many different species can inhabit the same pat. Many species in the same family (Hydrophilidae) are aquatic and the smooth lines and broad short legs of *Cercyon* specimens allows them to push right into soft dung.

Minute dung beetle, *Ootypus globosus*

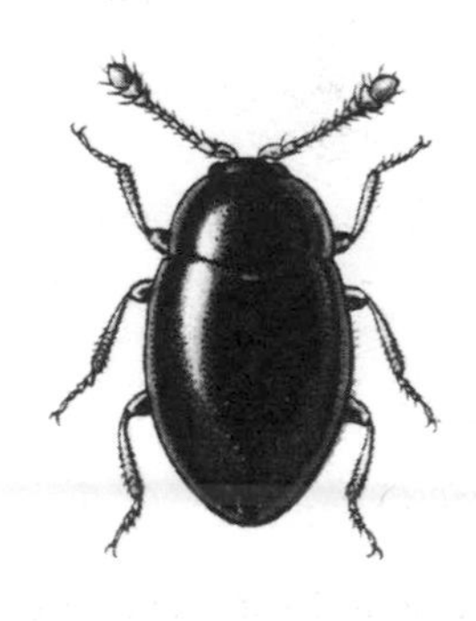

Size: 1.0–1.5 mm.

Description: Minute, convex, domed, almost hemispherical, very shiny beetle; black to dark reddish chestnut brown. Legs short and slim.

Life history: Little is known other than it breeds in dung and other rancid decaying organic matter. Several extremely similar species, including *Atomaria*, also named for their small size.

Ecological notes: Just about the smallest dung beetle to be found. Probably a fungus feeder, eating the fungal hyphae and grazing mould and spores.

Clown beetle, *Hister unicolor* (and other species)

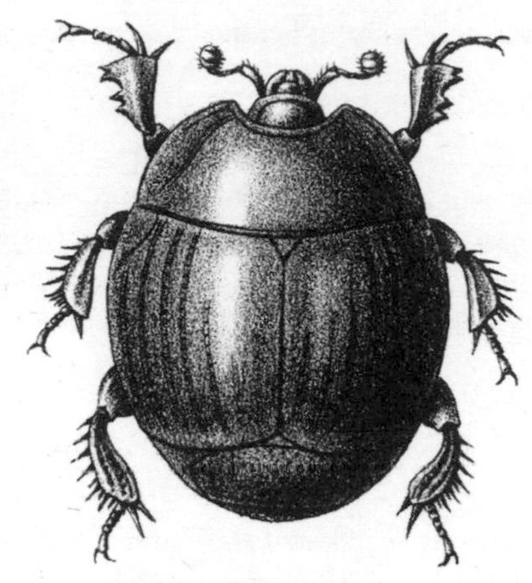

Size: Length 8 mm.

Description: Broad, round, domed, smooth, shining black, heavily armoured beetle. Antennae short, clubbed. Legs broad, armed with teeth, flattened, and which can be pulled in and recessed into grooves on the underside of the body. Many similar species, some marked with vague red blotches.

Life history: Adults and larvae are predatory on other dung-inhabiting invertebrate larvae, notably fly maggots.

Ecological notes: Not common, so play only a minor role in dung ecology, but distinctive and attractive beetles. Many similar species, some marked with reddish blotches, some metallic greenish or bronze. Also found in carrion, rotting fungi, decaying plant material. Sadly the obviously dung-named *Margarinotus merdarius* is mostly a species of compost and manure heaps, or bird nests in tree hollows. Smaller and shinier *Hypocaccus* species usually confined to sand dunes, where dog dung is only available food. Common name is of dubious etymology, either Latin *hister* meaning dirty lowly being, or *histrio* (from which also histrionics) meaning actor.

Ridged clown beetle, *Onthophilus striatus*

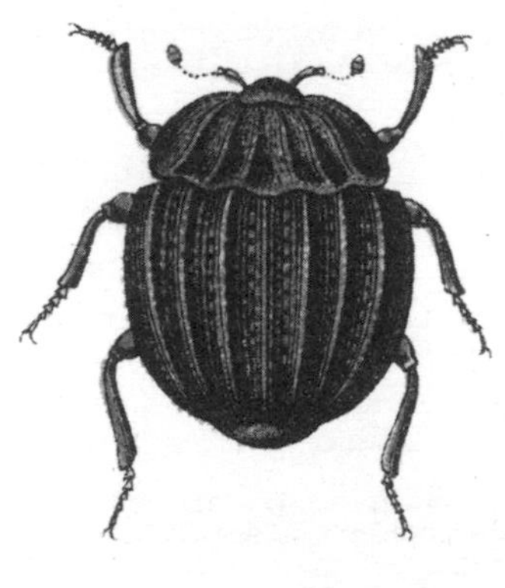

Size: Length 1.8–2.4 mm.

Description: Tiny, globular, dull black beetle. Wing cases each with three strong and three smaller raised ridges. Thorax with six raised ridges. Legs and antennae slim.

Life history: Predator of other even smaller dung invertebrates, particularly fly larvae.

Ecological notes: Seemingly only ever in horse dung, but closely related, larger (2.5–3.5 mm) *O. punctatus*, with five thorax ridges, in mole nests, fox dens or badger setts.

Feather-winged beetles, family Ptiliidae

Size: 0.5–1.2 mm.

Description: Microscopic black or brown beetles, short oval, some with wing-cases shortened, exposing the tail of the abdomen. Antennae long and very thin with rings of long hairs.

Life history: Feeders on fungal hyphae and spores.

Ecological notes: Found in old dung where powdery mildew fungi and moulds have started to grow. Large number of extremely similar species. Also found in leaf litter, rotten wood, fungi, animal nests, compost, mouldy hay, ant nests, decayed seaweed. Ironically named *Nephanes titan* (0.55–0.65 mm) one of many frequently described by entomologists in terms of 'the size of the full-stop at the end of this sentence'. Dung associations broad, but sometimes blurred; *Euryptilium gillmeisteri* known (in UK) only from three specimens found in leaf litter infected with bird droppings at base of large oak tree.

Ridged rove beetle, *Micropeplus porcatus*

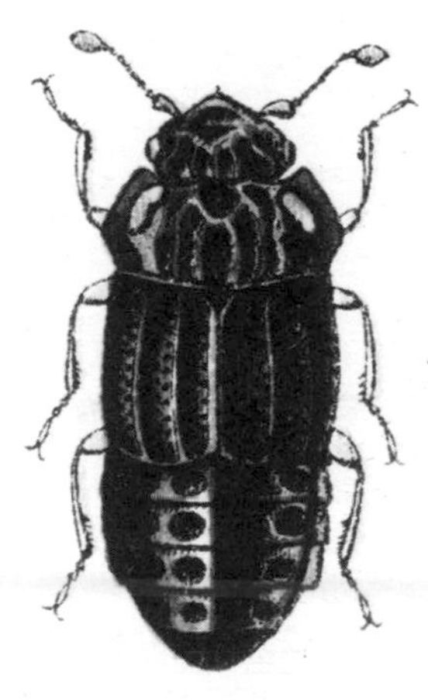

Size: Length 2.5–3.0 mm.

Description: Short, broad, flat, dull brown rove beetle. Wing cases each bearing three sharp ridges, short, exposing four or five abdominal segments. Thorax strongly sculptured. Legs and antennae very slim. Slow moving. Several very similar species.

Life history: Feeds on mould and fungal hyphae in decaying organic material.

Ecological notes: In old dung, also in manure heaps, rotten straw, mouldy hay, compost, leaf litter and other dry, decaying organic matter. Also occurs in mud at the side of lakes and streams, suggesting it feeds on liquid decay as well as dry fungoid rot.

Impossible rove beetles, *Atheta* species

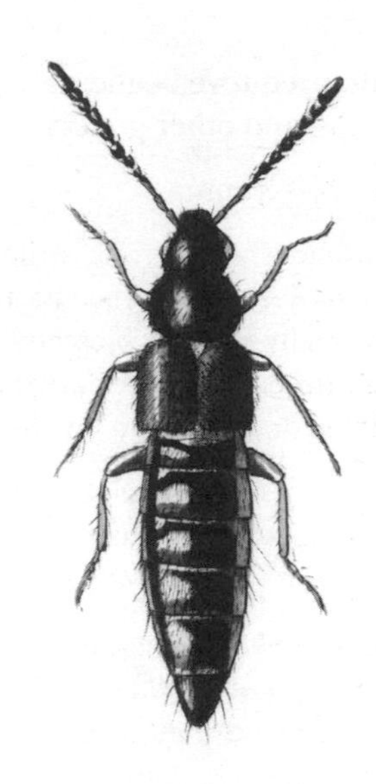

Size: Length 1.1–5.0 mm.

Description: Minute to very small, slim, black or brown rove beetles. Legs and antennae slim.

Life history: Very little is known about their habits. Some species are known to eat free-living nematode worms, so predation of micro-organisms and other soft-bodied invertebrates seems likely.

Ecological notes: Occur in a wide variety of decaying organic matter with several specifically in dung. Huge number of extremely similar species, which require expert knowledge and detailed examination to identify. Most identifications rely on dissection and special mounting of internal genitalia, notably the female's sperm-storage organ (spermathecal). Males of some species unknown or not possible to identify. That's not really their English name, just one I have coined in my own frustration at finding them so impossible to identify correctly.

Flat rove beetles, *Megarthrus* species

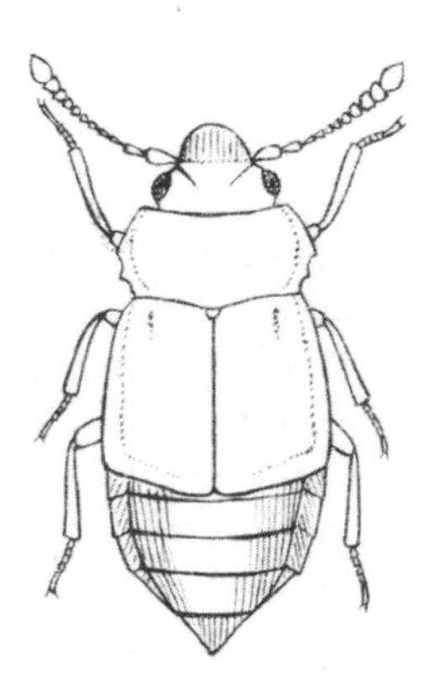

Size: Length 2.5–3.5 mm.

Description: Short, very broad, flat, brown or black rove beetles. Hind corners of thorax incised with small notch. Wing-cases long (for a rove beetle), about as long as, or slightly longer than hind body.

Life history: Both adults and larvae are predators of smaller invertebrates.

Ecological notes: In dung, manure or other decaying organic matter. Many similar genera.

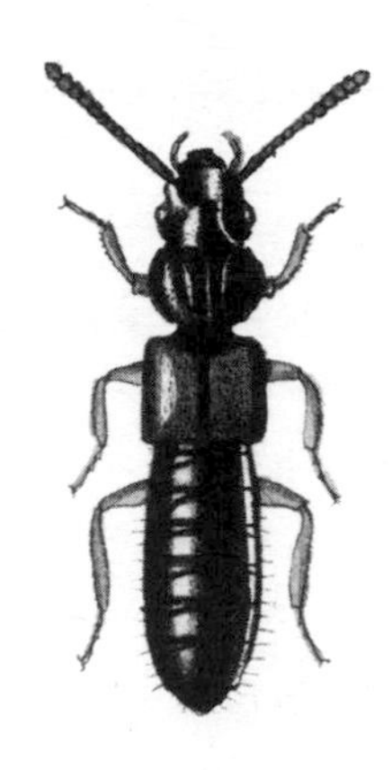

Sculptured rove beetles, *Platystethus, Anotylus, Oxytelus* and other genera

Size: 1.2–5.5 mm.

Description: Numerous group of slightly flattened, black or dark brown rove beetles, sometimes smudged with pitchy-reddish blotches. Very parallel-sided with broad heads, broad rectangular thorax usually ridged, dimpled or grooved to give sculptured appearance.

Life history: All feed as larvae in decaying organic matter, including dung, manure, compost, rotting seaweed, putrescent fungi, haystack refuse, leaf litter, and moss.

Ecological notes: Slow-moving under dung. Spiny legs imply they burrow in the dung or soil beneath. Other closely related groups with similar broad, shovel-shaped front legs are confirmed soil-diggers.

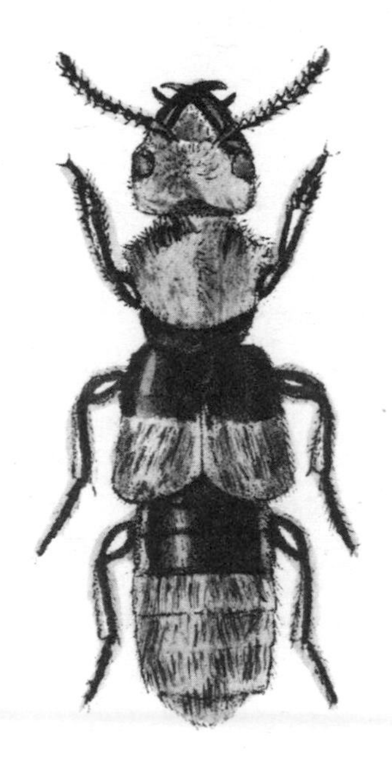

Pride of Kent, *Emus hirtus*

Size: Length 18–28 mm.

Description: Large, broad, powerful, spectacular rove beetle. Black, but head, part of thorax and tail tip covered with bright yellow hairs. Rear half of wing-cases thick with grey fur. Legs short and stout. Jaws large.

Life history: Predator of other invertebrate inhabitants of cow dung.

Ecological notes: Extremely rare in the UK, with most records from the grazing marshes of north Kent. There was much excitement when it was rediscovered, after many years, running on the tarmac near a public lavatory on a Kent nature reserve. Usually associated with very fresh cow pats, on which it scurries in frantic activity, attacking the flies which are attracted. Possibly a bumblebee mimic when flying. Note: *Creophilus maxillosus* is another large, broad, furry rove beetle, but is patterned with pale grey hairs, not yellow; it occurs in carrion rather than dung, and is sometimes misidentified as *Emus* by excited, but inexperienced, observers.

Mottled rove beetle, ***Ontholestes tessellatus***

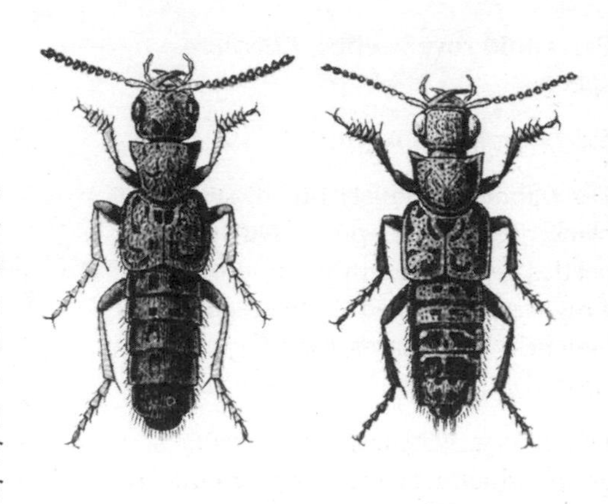

Size: Length 14–19 mm.

Description: Large, stout, rove beetle. Black, but forebody (head, thorax, wing-cases and first segment or two of abdomen) patterned with golden-green hairs to give a shifting chequered appearance. Small patch of matt black hairs at join of thorax and wing-cases. Tail black, marked with grey patches. Legs reddish golden. Smaller (10–15 mm) and narrower, *O. murinus* (right) is less golden, with legs black.

Life history: Predator of flies and their larvae. Also occurs on carrion.

Ecological notes: Very fast and active insect. Despite its silky appearance it copes with sticky dung very well, and grooms itself using front legs on head and antennae, middle legs for thorax and wing-cases, and rear legs for tail segments.

Red-girdled rove beetle, ***Platydracus stercorarius***

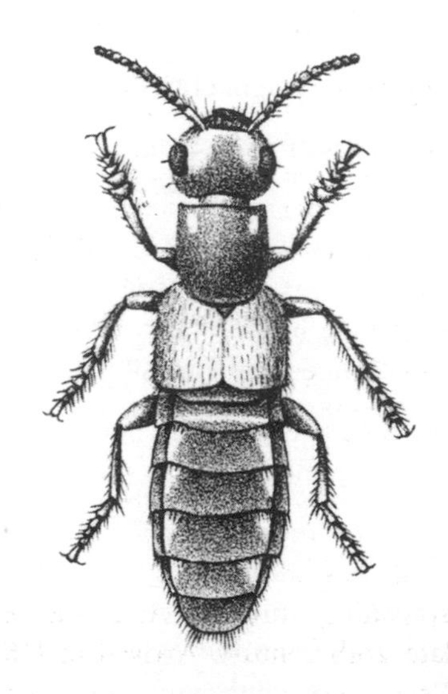

Size: Length 12–13 mm.

Description: Black, not very shining, but wing-cases strongly contrasting bright orange-red; abdominal segments decorated with bands of silver hair, which catch the light as the animal runs; legs and antenna red. Several similar species, some with wing-cases black, or all-over mottled with vague silvery grey pubescence.

Life history: Predatory as both larva and adult.

Ecological notes: Occurs under dung, but also under carrion, in manure and compost heaps. Very agile and active.

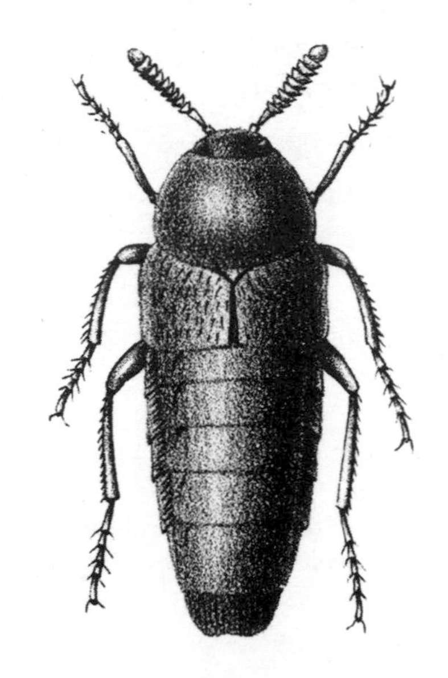

Parasitoid rove beetles, *Aleochara* species

Size: Length 1.5–10 mm.

Description: Relatively broad, stout, black, brown or pitchy-red rove beetles, usually shining; sometimes wing-cases marked with reddish blotches. Wing-cases seeming very short compared to long hind body.

Life history: Very large and diverse group, which are specialist predators of fly puparia; feeding completely inside the pupa, they may really be described as parasitoids.

Ecological notes: In all types of dung, but also in carrion and other putrescent decay. When fully grown, larva pupates inside the almost empty fly puparium. Adult beetle emerges through a jagged hole it chews in the puparium shell. This contrasts with the usually perfectly round hole through which parasitic wasps emerge.

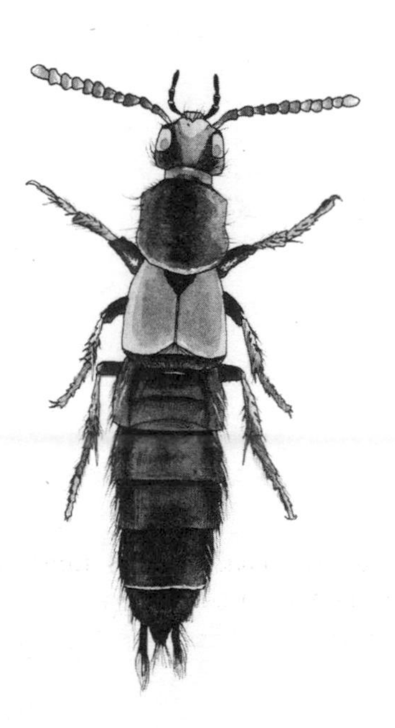

Japanese rove beetle, *Philonthus spinipes*

Size: Length 13–17 mm.

Description: Head and thorax shining black; wing-cases orange-red, covered with red pubescence; hind body iridescent black. Antennae black. Bristly legs black at base, and orange from knees.

Life history: Predator as both larva and adult. Very fast and active.

Ecological notes: Particularly in horse dung. First described (1874) from Japan, but spreading through Asia and Europe in late 20th century. Arrived in UK in 1997. Reason for range expansion unknown.

Splendid rove beetle, *Philonthus splendens*

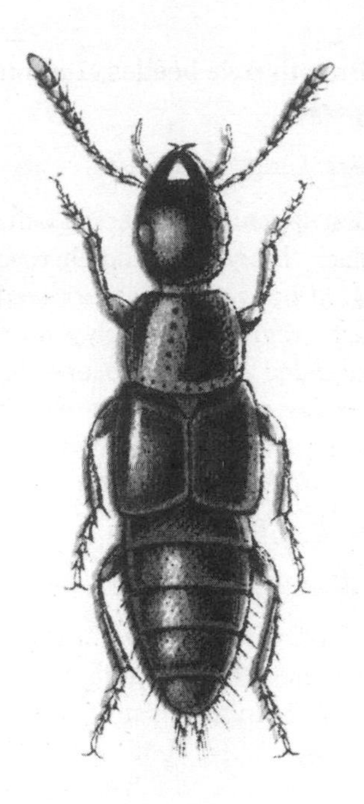

Size: Length 12–14 mm.

Description: Shining black parallel-sided rove beetle; head and thorax with slight metallic bronze sheen, wing-cases more strongly brassy green metallic.

Life history: Predator as both larva and adult, mostly on flies and their maggots.

Ecological notes: Very fast and active predator of other small invertebrates in dung, also manure heaps, carrion, putrid fungi. Very many similar species requiring expert knowledge to identify. Flies readily and can land, fold away wings in an instant and vanish under fresh dung in a blur.

Devil's coach-horse, *Ocypus olens*

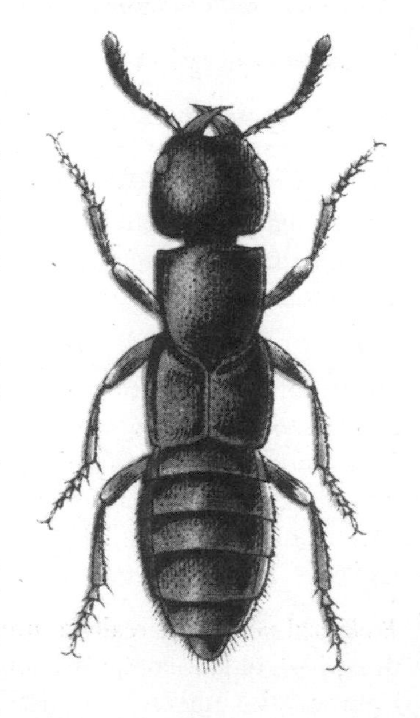

Size: Length 20–28 mm.

Description: Huge, broad, dull black rove beetle. Entire body surface covered with dense array of minute pinprick dints and short black pubescence, giving a matt appearance.

Life history: Predator as both larva and adult. Commonly under stones, logs, clods of earth, in compost and manure heaps. Not a specialist dung species, but regularly found treating old dung as suitable shelter.

Ecological notes: Often found running on paths, and familiar to gardeners and many non-entomologists. If threatened will rear up head, displaying its large jaws (can nip), and tail, exuding droplet of smelly liquid from its tip.

Smooth rove beetles, *Tachinus* species

Size: Length 3–10 mm.

Description: Elegant, smooth, shining black, brown or reddish rove beetles. Hind body strongly narrowed behind, last segment deeply notched, or extended at sides into long teeth. Many similar species. Note: closely related and hugely abundant genus *Tachyporus* specimens are smaller (2–4 mm), and occur in grass thatch.

Life history: Predator as larva and adult.

Ecological notes: In rancid decaying matter such as carrion, manure, putrid fungi, compost heaps and dung.

Burying beetles, *Nicrophorus* (formerly *Necrophorus*) species

Size: Length 10–30 mm.

Description: Large, stout, heavily built, parallel-sided beetles, generally black with bright orange bars (broken or solid) across wing-cases; some species all black, others orange on thorax.

Life history: Carrion feeders, mostly of small carcasses such as voles or birds. Dig out earth from underneath creating a void into which the body subsides, and spoil from which eventually engulfs it. Then feeds on putrescent decay, and lays eggs. Male and female generally work together.

Ecological notes: Not really a dung feeder, but often found under carnivore dung if stool still contains enough decaying meat material to give off the right scents. Unlikely to result in interment, egg-laying or breeding though.

Hide beetles, *Trox scaber* (and other species)

Size: Length 5–7 mm.

Description: Stout, blunt, oval, domed beetle, body roughly sculpted with raised tubercles and covered with short upstanding scales or broad stiff hairs, giving a slight bristly appearance, and sometimes obscuring the beetle by accumulating a covering of dust and debris.

Life history: Mostly carrion feeders, appearing at the dry tendon-and-fur stage after putrescence has passed.

Ecological notes: Not really a dung feeder, but sometimes found under carnivore scats, especially if they are rich in fur, feathers or hair.

Dumbledor or dor beetle, *Geotrupes stercorarius* (and other species)

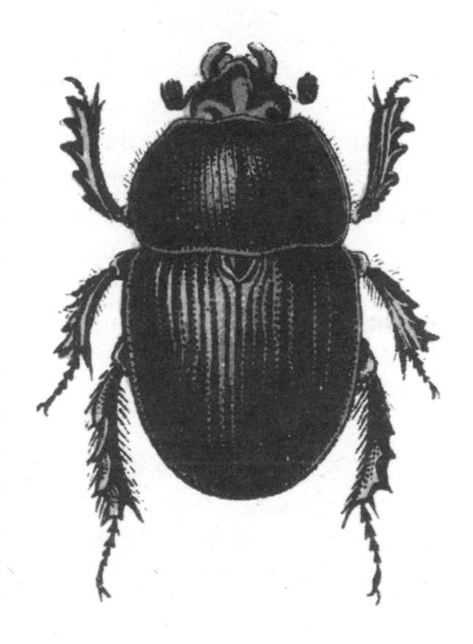

Size: Body length 16–26 mm.

Description: Massive, convex, broad domed, shining black beetle, with purplish or bluish tints, stout powerful toothed legs, antennae clubbed.

Life history: Adults dig deep burrows into the soil, provisioning side tunnels with dung balls in which eggs are laid. Fat, pale whitish C-shaped grubs feed in these dung cells and emerge as adults, pushing up through soil, weeks or months later (depending on temperature). Some species also use decaying leaf litter instead of dung.

Ecological notes: Important dung recyclers because of their size. Despite heavy build, they are good fliers across meadows on warm evenings, giving off loud buzzing whirr, leading to the onomatopoeic English names; also known as lousy watchman (often found carrying louse-like mites, watchman is reference to 'clock', an old name for any large beetle). Several different species distinguished by size, colour hints, leg tooth shape and smoothness or ridge numbers on wing cases.

Minotaur beetle, *Typhaeus typhoeus*

Size: Body length 14–22 mm.

Description: Stout, convex, broad domed, shining black beetle with broad, toothed legs and clubbed antennae. Male with three sharp spines projecting over head from front of thorax, usually side ones longer, but varying greatly; female with slight bumps instead.

Life history: Like dor beetle, buries dung balls in a tunnel dug down in the soil, on which eggs are laid.

Ecological notes: Prefers sandy soil where burrows can be up to 2 m long; mostly on rabbit or sheep dung. Drags, rather than rolls, the small pellets. Overwinters as an adult in the burrow so is sometimes active on mild winter days. Attracted to lighted windows and porches, and regularly found in moth light traps, even in early February.

Lesser scarab, *Odonteus armiger* (formerly *Odontaeus mobilicornis*)

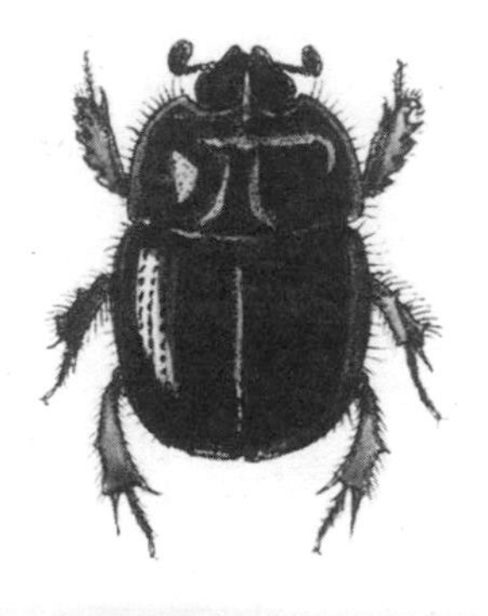

Size: Body length 6–8 mm.

Description: Short, round, almost globular, smooth, shining, dark brownish black, legs reddish. Male with long backwards curving moveable spine on its head, female with small prominences.

Life history: Usually found flying on warm summer evenings. Thought to feed on subterranean fungi, and excavates burrows to 40 cm deep.

Ecological notes: Reported from dung in some of the old books, and possibly associated with buried dung on which fungi develop. Closer association with rabbit warrens has been suggested, but evidence is flimsy.

Common dwelling dung beetles, *Aphodius* species

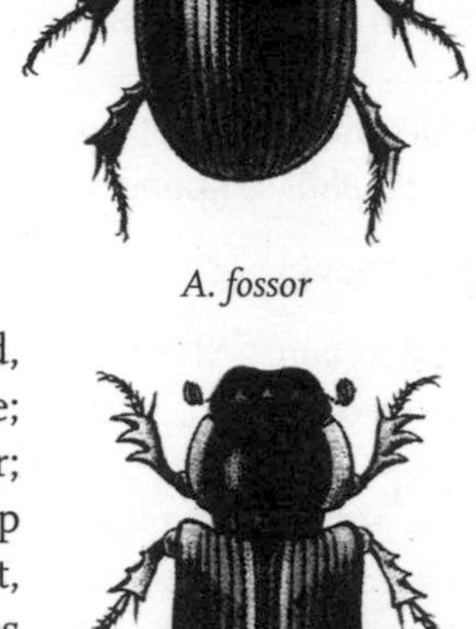

A. fossor

Size: Length 2.5–15 mm.

Description: Stout, semi-cylindrical long oval beetles, with wing-cases distinctively rounded at tail end, broad domed thorax and blunt shovel-shaped head. Without horns or prominences on head or thorax. Very many different species. Typical common European species include: *A. fimetarius* (6–8 mm), jet black but wing-cases and front corners of thorax bright red; *A. contaminatus* (5–6 mm) wing-cases straw coloured, flecked with black marks in vague chevron shape; *A. sticticus* similar but shinier, markings crisper; *A. rufipes* (11–13 mm) long, parallel-sided, deep chestnut red-brown; *A. fossor* (9–12 mm) thickset, black, shining; *A. prodromus* (4–7 mm) wing-cases dirty yellow, with large darker smudge at sides and end; *A. erraticus* (6–9 mm) wing-cases dull beige, very broad; *A. haemorrhoidalis* (3–4 mm) all black, wing-cases with reddish tip.

A. prodromus

Life history: Fly readily to colonise dung, burrow into and beneath the dropping, where they lay their eggs. Huge numbers and/or multiple species can occur in a single pat.

Ecological notes: In temperate regions this is the most important and numerous dung beetle genus. Despite their ubiquity they have no widely accepted common name, so

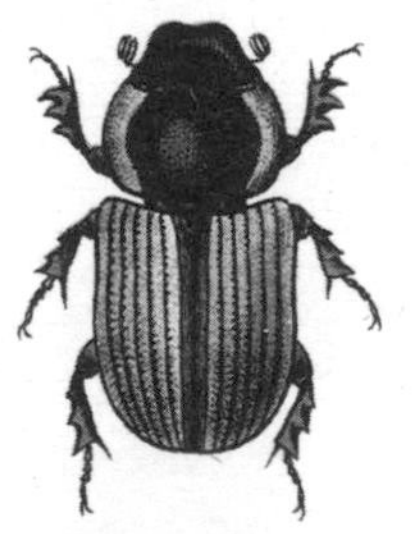

A. nitidulus

A. varians

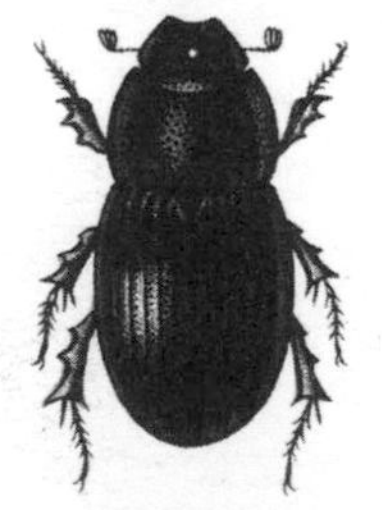

A. haemorrhoidalis

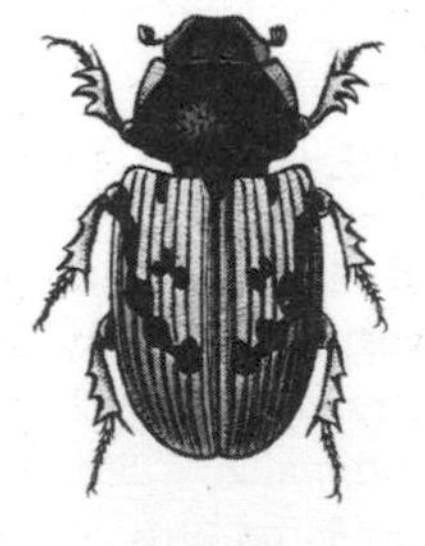

A. sticticus

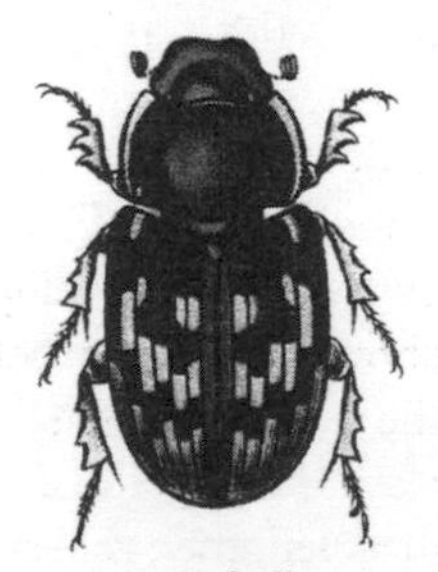

A. paykulli

I've coined this one since they are the dominant 'dweller' beetles (as opposed to tunnellers or rollers). Some species supposedly more associated with dung of particular animals. There is some evidence that moisture content is important so some species cope better with wet cow dung, others are more at home in drier sheep or rabbit pellets, but most are fairly broad in tastes; thus deer dung species more likely to be just adapted to woodland shade rather than stool content.

Common tunnelling dung beetles, *Onthophagus* species

Size: Length 3–14 mm.

Description: Short, compact, oval convex beetles, head and thorax about equal in size to abdomen covered with wing-cases. Legs flat and broad, armed with large teeth. Variously coloured black, yellowish, brown, reddish; wing-cases sometimes mottled or blotched, matt to polished shining black, metallic green or bronze. Males often with pronounced, sometimes bizarre, horns and spines from head or front of thorax. Widespread (but never common) European species include: *O. joannae* (4–6 mm) all dull black; *O. taurus* above right (8–12 mm) all shining black or dark brown, male head with two long curved horns reaching back over sides of thorax; *O. coenobita* (6–10 mm) head and thorax dull dark metallic green, wing-cases pale brown, usually flecked with vague darker blotches; *O. similis* bottom (4–7 mm) thorax black, wing-cases dark beige with chequer-like pattern of black speckles.

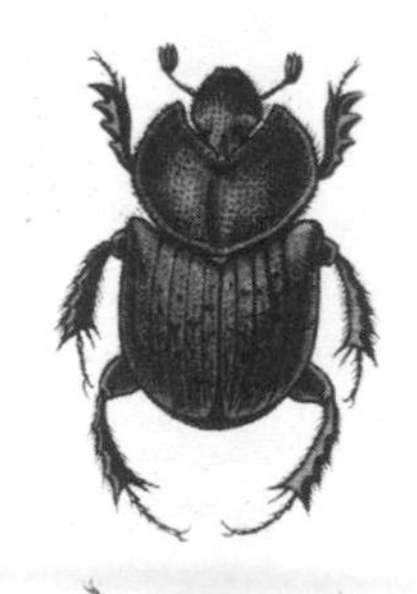

Life history: Digs burrow under the dung and removes pellets or boluses down into the blind end, on which to lay eggs.

Ecological notes: Hugely varied and important genus of beetles, and the dominant tunnelling genus through most of the world. Some of the horns are truly spectacular and the basis of studies trying to understand how and why these decorations have evolved and developed. Head-to-head pushing and shoving contests between males may take place in the tunnels. Three Australian species have prehensile claws to grip fur around wallaby anus until dung is dropped, then release hold and bury pellet whole.

English scarab, *Copris lunaris*

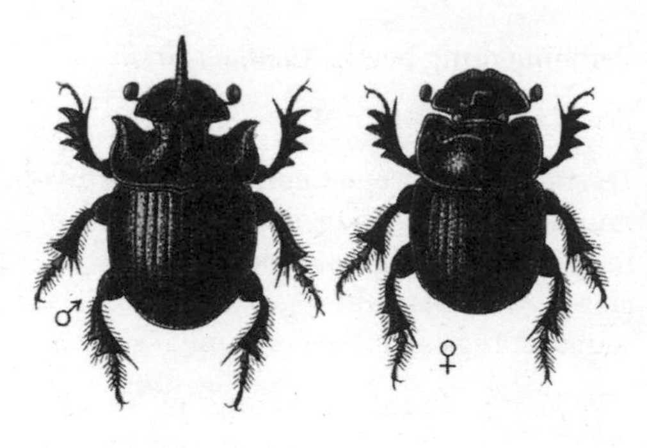

Size: Body length 17–23 mm.

Description: Stout, domed, shining black beetle, but with sharp ridge along cut-in front edge of thorax, and head with long (male) or short (female) spine projecting up from broad, rounded head.

Life history: Like dor beetle, buries dung balls in a tunnel dug in the soil, on which eggs are laid.

Ecological notes: Usually on cow dung and said to prefer sandy or chalky places. Extremely rare in Britain, known in UK only from a handful of sites in south-east England and likely to be extinct; not reliably recorded since 1950s. 'English' scarab after conservation exercise in 1990s meant priority species were all given common names. Widespread and sometimes common in Europe.

Tumblebugs, *Canthon pilularius* (formerly *C. laevis*) and other species

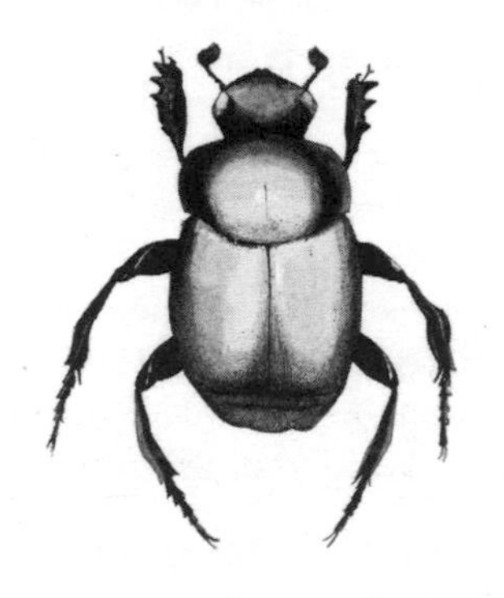

Size: Length 10–20 mm.

Description: North America. Broad, smooth, dull black beetle, with coppery tinge or (south-western colonies) brighter blue or green sheen. Head broad, strongly rounded and flattened. Front legs broadened with three deep teeth; middle and back legs longer and slimmer.

Life history: Flies in low zigzags upwind to fresh dung. Landing, it immediately starts to cut away a lump of dung to shape it into a ball. Balls are roughly 30 mm across and take about 20 minutes to sculpt. Sometimes the ball is rotated in the dry ground to give it a coating of sand. The ball is then rolled away from the pat, to be buried.

Ecological notes: Rolling is by the beetle pushing, head down, standing on its front legs and using its middle and back legs to manoeuvre the ball away. Quite an ungainly animal, by all accounts, frequently tumbling and losing its grip on the ball, which rolls away and has to be searched for, antennae outstretched. Route away from the pat not very straight, often zigzag.

Perching dung beetle, *Canthon viridis*

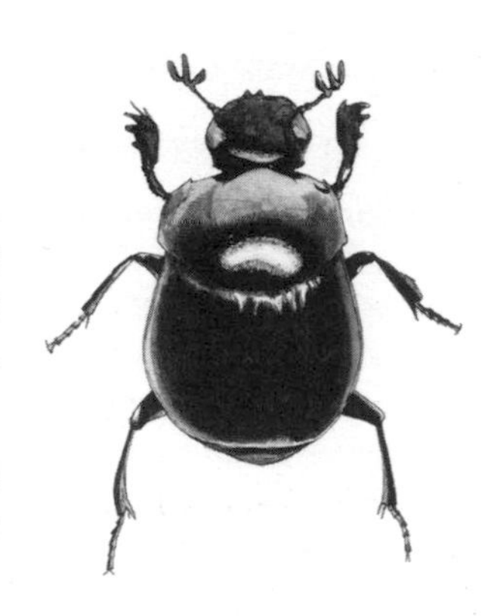

Size: Length 2–5 mm.

Description: North and Central America. Small, round, stout, globular, dung beetle; shining metallic purple, bronze (northern) or bright glossy green (southern). Head broad, flat, rounded. Legs slim. Several similar species.

Life history: Breeds in dung of woodland animals such as deer, monkeys, peccaries; also in rotting fungi and carrion.

Ecological notes: A woodland or forest species. Noted for its perching behaviour, on leaves and stems in the dappled light of often dense treescapes, its round little body upright on its slim legs and with its antennae outstretched, three flat terminal club segments fanned out. Possibly waiting to catch scent of freshly dropped dung, or waiting for dangerous predatory rove beetles to leave the dung.

Carolina scarab, *Dichotomius carolinus*

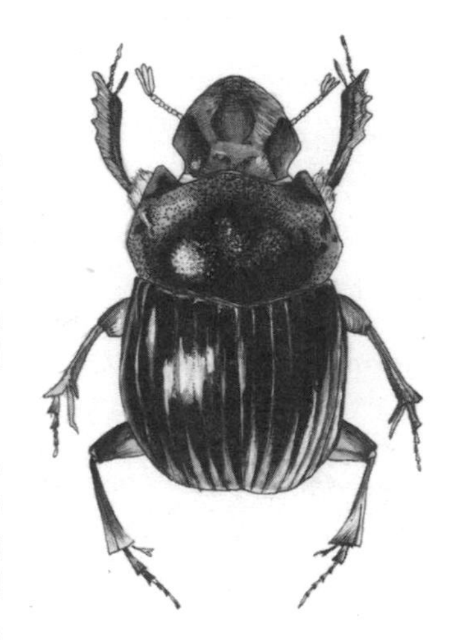

Size: 20–30 mm.

Description: North America. Large, heavy, broad, stout dung beetle; shining black, but underside fringed with brown or orange hairs. Legs black, antennae orange-brown. Wing-cases deeply grooved. Thorax with heavy jutting shoulder across front. Head large, broad, rounded, almost flanged.

Life history: Breeds in dung of horses, cows, deer, digging a tunnel 30–40 cm deep into the soil and packing its end with dung fragments to make a food store or brood ball. Single white egg laid on brood mass. Generation time about 2 months. Large mounds of excavated soil mark the burrow entrances.

Ecological notes: Nocturnal, and often attracted to lighted porches or verandas during the summer. Thought originally a forest species, of open clearings, but regularly found in grazing pastures abutting woodland. One of the largest and heaviest dung beetles in North America. So powerful it is difficult to hold in a clenched fist.

Rainbow scarab, *Phanaeus vindex*

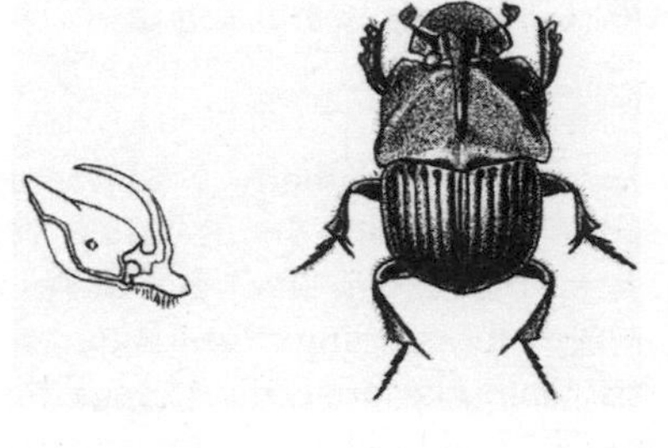

Size: Length 11–22 mm.

Description: North America. Beautiful and striking brilliant metallic green and gold, sometimes blue or black, dung beetle; broad, almost square. Thorax flattened and produced into two broad backward sloping, tooth-like lobes. Head of male armed with huge, backwardly curved horn. Whole body covered in wrinkled, dimpled sculpture. Legs broad and stout. Several similar species.

Life history: Tunneller, digging burrow under the dung and throwing up spoil heap beside the pat. Roughly pear-shaped dung ball is placed at the end of the burrow. Side tunnels may be created to accommodate more brood masses. Single egg on each mass.

Ecological notes: Widespread and often common; in wide variety of dung including pig, opossum, dog, cow, horse and human.

Giant African dung beetle, *Heliocopris gigas*

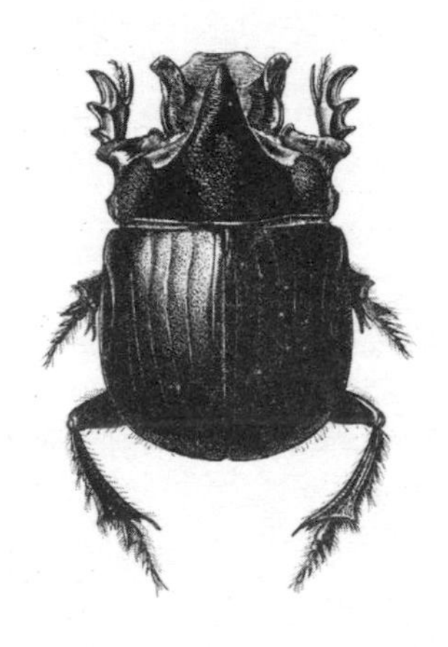

Size: Length 37–60 mm.

Description: Africa. Very large and stout, domed, broad, almost square black or very dark brown dung beetle. Black, slightly shining, underside with fringing of brown hairs. Legs broad, especially front pair, flattened and armed with large teeth. Head very large, broad, round, spade-shaped in female, armed with two broad twisted back-swept horns in male. Thorax with nearly vertical cliff-like front, produced, in males, into sharp spines at sides and a large prong jutting over the head. All over dimpled and wrinkled to give tough leather-like appearance

Life history: Breeds in elephant and other dungs, digging a tunnel beneath the pat and stocking the ends of the burrows with brood balls, each about 5 cm in diameter, on which single eggs are laid.

Ecological notes: Although once thought to breed only in elephant dung, it occurs widely in the elephant-free Arabian Peninsula, and will use droppings from wild animals, also camel, cow and horse.

Sacred scarab, *Scarabaeus sacer*

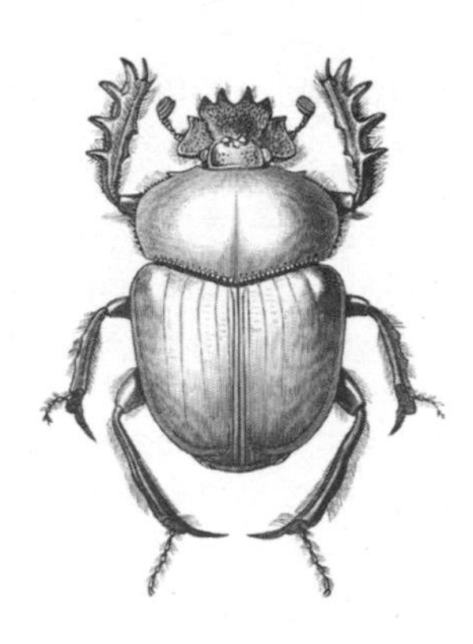

Size: Length 26–40 mm.

Description: Southern Mediterranean Europe and North Africa, Indus Valley. Very large and stout, broad, flat, short oval dung beetle. Black, moderately shining. Head flattened, armed with six tooth-like projections. Legs relatively long; middle and hind pairs slim, front pair with each leg armed with four broad teeth. Several similar species.

Life history: Scoops balls of dung using its broad head and rake-like front legs, then rolls these balls backwards, head down, away from the pat some distance (often several metres) before burying them in the soil. Each brood ball with a single egg laid in it.

Ecological notes: Uses the sun (or moon, or indeed Milky Way) to orient its journey so that even though it does not look where it is going, it can successfully roll away the dung ball in a straight line, despite uneven ground, ridges, boulders or log barriers which would otherwise interrupt its path.

Egyptian scarab, *Kheper aegyptiorum* (and other species)

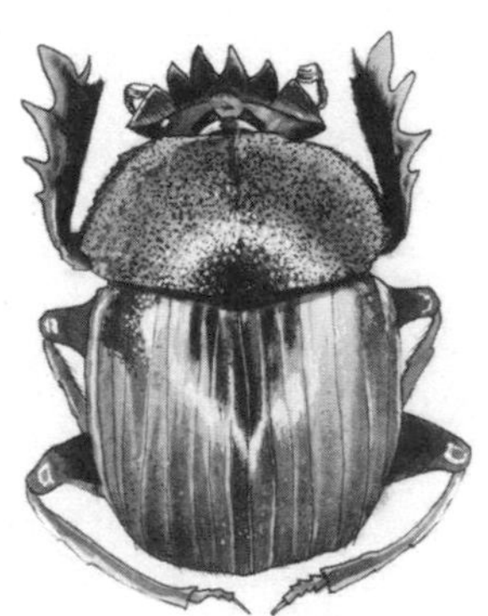

Size: 25–40 mm.

Description: North-east Africa. Very large, broad, flattened, short oval dung beetle. Black, moderately shining, often with vague metallic bronze tinge. Flat head armed with six projections. Front legs similarly armed with flat blade-like teeth.

Life history: Like *Scarabaeus* (it is sometimes also included in this genus) rolls large ball of dung away from the main pat, and buries these in the soil some distance away.

Ecological notes: Active during the day, where many African dung beetles are semi-nocturnal. Once a male has shaped a dung ball and moved it a short distance from the main pat, it adopts a head-stand pose, angled at about 45°, on top. It is releasing a nuptial pheromone from the underside of its abdomen, pumped into the air by simultaneous contractions of its back legs. This attracts a female and the two beetles work together to bury the dung ball, before the female lays an egg in it.

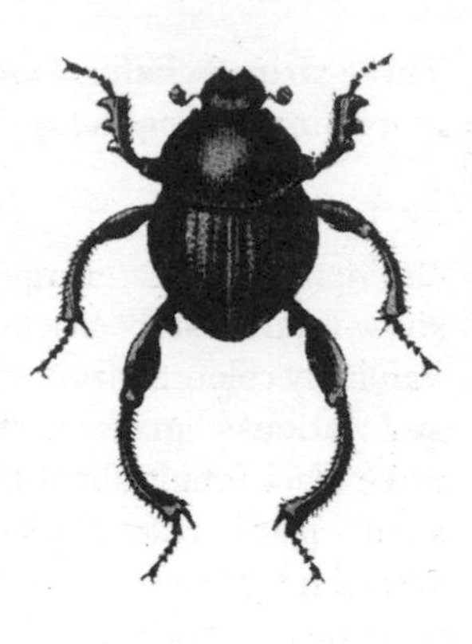

Gracile dung beetle, *Sisyphus* (now sometimes *Neosisyphus*) *mirabilis*

Size: Length to 12 mm.

Description: South Africa. Body small oval, dull black; head broad, rounded, paddle-shaped. Legs very long, especially hind pair which are armed with various spines and pegs. Middle legs each armed with strong curved spine. Front legs short, stout, armed with small teeth. Several similar species worldwide, though not all as long-legged.

Life history: A dung roller, digging out a marble-sized ball of dung and rolling it off at high speed away from the pat, to a safe place where it can be buried in the soil and an egg laid.

Ecological notes: It rolls by running on its front legs, head down, tail in the air, using its hooked middle legs and long back legs to keep control of the ball it is pushing. In Greek mythology, Sisyphus was the conceited king of Ephyra (Corinth), condemned to an eternity of rolling a large boulder up a hill, only to have it roll away from him near the top so he would have to start all over again.

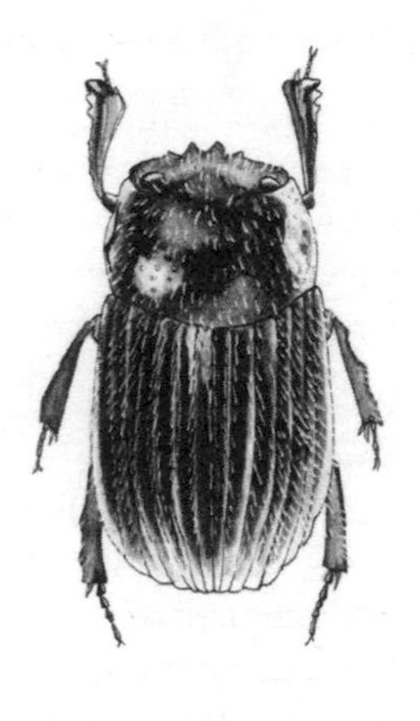

Sloth dung beetle, *Pedaridium* (formerly *Trichillum*) *bradyporum*

Size: Length 3 mm.

Description: South and Central America. Small short oval, domed, very convex dung beetle. Black or pitchy with reddish brown tints, verging on metallic. Upper surface covered all over with short, widely spaced, but distinct upstanding hairs. Legs short.

Life history: Associated with three-toed sloths (*Bradypus* species), in the fur of which they lodge (nearly 1,000 recorded from one individual). When the host descends to the forest floor to defecate (they bury their dung to hide their presence from predators), the beetles drop off and lay their eggs.

Ecological notes: Several similar species are recorded from cow and human dung, but close association with any particular hosts is unknown or unrecorded.

Three-striped chafer, *Macroma* (sometimes *Campsiura*) *trivittata*

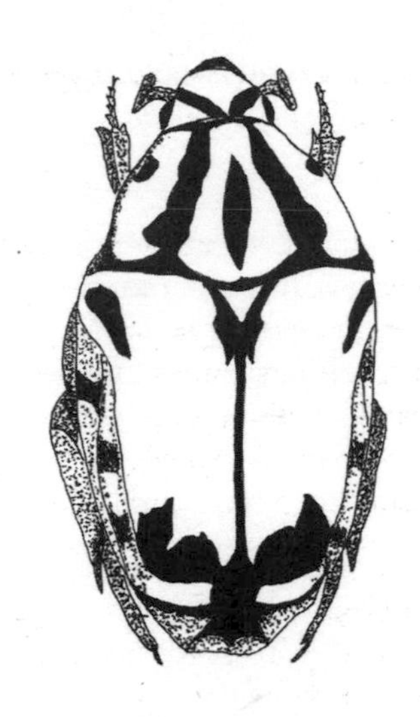

Size: 20–23 mm.

Description: Africa. Large broad, smooth straw-yellow and black chafer beetle. Varyingly coloured from yellow to reddish, and variously marked, but usually with three black longitudinal stripes on thorax, small black tick marks on shoulders and jagged black bar across near tip of wing-cases. Legs and antennae reddish.

Life history: Fat C-shaped grub lives in leaf-litter and soil, eating decaying plant material. Adults fly actively.

Ecological notes: Has been recorded in Ivory Coast breeding in elephant dung. Unlike in the open savannahs, there appears to be less competition for dung in the tropical rainforest hereabouts and elephant dung remains in large quantities for secondary users.

Cave larder beetle, *Dermestes carnivorus*

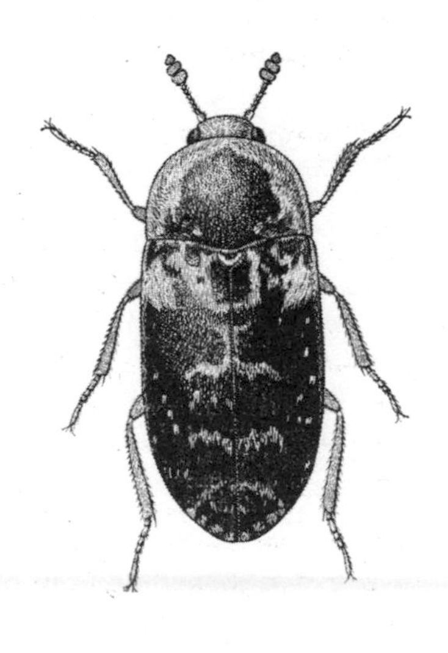

Size: length 6.5–7.5 mm.

Description: Small, long-oval beetle. Black, upperside with scattering of brownish scales, underside with dense blanket of silver-white scale-like hairs. Legs medium length. Antennae brown, clubbed.

Life history: Active bristly larvae are scavengers. Original habitat probably carrion, at its dried stage where only bone, sinew, fur and feather remain. Thought to have originated in neotropics, but after invading human homes as larder pest (in stored meat products) has been transported across the globe.

Ecological notes: Major decomposer of bat guano in caves. Will probably also feed on dead bats. As well as making jump to human habitations to feed on spilled food, has been recorded in bat droppings in abandoned buildings.

Shining spider beetle, *Gibbium aequinoctiale*

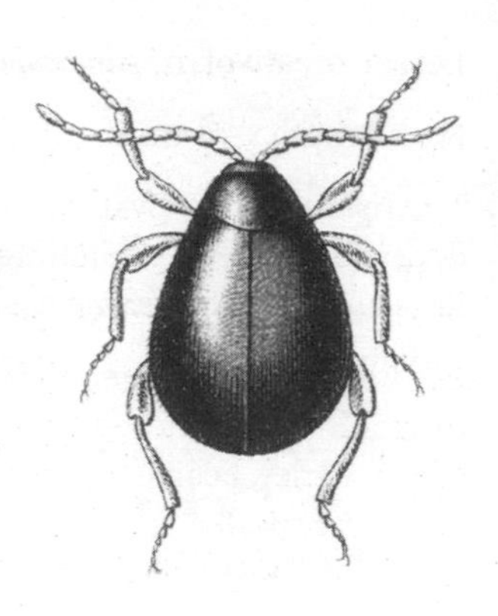

Size: Length 1.5–3.5 mm.

Description: Small, shining globular beetle. Brown to beige, wing-cases usually glossy chestnut. Abdomen nearly spherical, humped, head hidden under narrow thorax. Legs relatively long, giving spider-like appearance.

Life history: A scavenger species, part of a large group of similar species that feed on spilled food in houses and warehouses, or in seeds and nuts cached by wild animals or dropped inside their nests.

Ecological notes: Completely bizarrely found 800 m underground in Yorkshire coal-mines, feeding on human dung. No toilet facilities being available in the mine workings, miners allocate disused tunnels as unofficial latrines. With no 'natural' dung fauna to bury or recycle it, the excrement remains, but dries out into a cake-like consistency. The beetle is flightless, lacking wings, and with its wing-cases fused together; how it got down the mine (and others in Durham and Staffordshire) is still a mystery.

Click beetles, *Agriotes* species

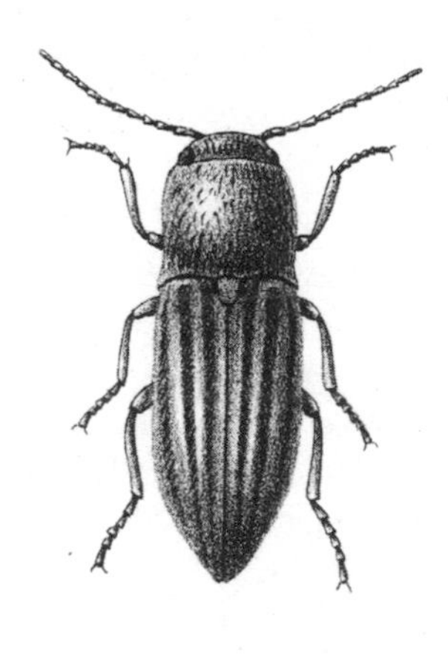

Size: Length 4–9 mm.

Description: Small, smooth-outlined, long-oval parallel-sided, almost cylindrical beetles; brown, greyish black to dull chestnut. Legs slim, antennae long. Many similar species.

Life history: Soil-dwelling larvae (called wireworms by gardeners) are long, cylindrical, pale yellowish-orange with brighter orange, brown or black head capsule. They feed in the soil, partially predatory, but also eat plant roots, and are sometimes regarded as horticultural or agricultural pests.

Ecological notes: Not really dung feeders, but larvae regularly invade the blurred soil horizon under old dung; adults less so. Adult beetle can make audible click by sudden jackknifing of body, which propels it into the air, or out of harm's way, if attacked or disturbed.

Lesser mealworm, *Alphitobius diaperinus*

Size: Length 5–6 mm.

Description: Long oval, smooth, glossy, black or dark brown beetle, sometimes with reddish tinge. Legs and antennae same colour as body.

Life history: A scavenger of dry material, including fungoid wood, stored food products such as wheat, barley, beans, tobacco and dried meat. Has been transported around the globe and can be serious domestic or commercial pest.

Ecological notes: Often found in caves, feeding primarily on dry bat guano. It is also found in chicken houses, where it may be feeding on droppings.

Dung weevil, *Tentegia ingrata*

Size: Length 8–12 mm.

Description: Australia. Dumpy, hunched weevil with bulbous thorax, wing-cases ridged or knobbled, dark, sometimes black, sparsely flecked with small paler scales. Snout long and narrow. Legs relatively long.

Life history: Adult beetle awkwardly manoeuvres pellets of wallaby or kangaroo dung from dry grassland several metres to small log, under which it caches them. It makes a small hole in the soil, or uses already existing cavity, to store the dropping, then lays an egg in each pellet. Very similar *T. bisignata* has been reared from pellets of possum dung found hidden in similar caches. It is unclear why the unfortunately named *T. stupida* is so called.

Ecological notes: The only known dung-feeding genus of weevil. Australian marsupial droppings are dry, so the weevils treat it as merely preprocessed plant material. Ordinarily female weevils use their long snout, with small biting jaws right at the end, to chew a deep drill hole into leaf, stem, seed, then turn round and use a telescopic ovipositor to lay egg deep inside. In this case it seems that they chew into the dung nugget to make an egg tunnel.

LEPIDOPTERA – BUTTERFLIES AND MOTHS

Butterflies, various species

Size: Wingspan 25–105 mm.

Description: Unmistakable. Distinctive colourful day-flying insects. Huge variety of different sizes, shapes, colours and patterns.

Life history: Mostly plant-feeders as caterpillars (with chewing mouthparts), taking weeks or months to achieve bulk enough to pupate and transform into adult. Adults have long tongue for sipping nectar from flowers.

Ecological notes: Not really dung feeders, but regular visitors (as adults) to fresh droppings where they sit and drink from the moist surface. In tropical countries muddy banks of streams and ponds, where animals drink and defecate, can be ablaze with scores, even hundreds, of butterflies, often a mixture of different species. This 'puddling' behaviour likely supplies minerals, salts and other micronutrients.

Sloth moth, *Cryptoses choloepi* (and other species)

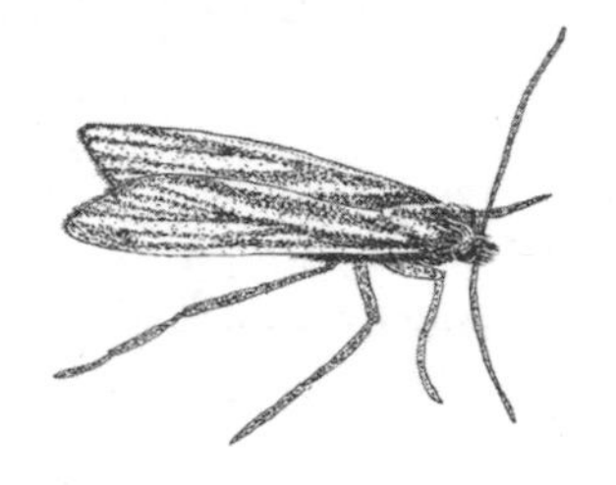

Size: Wing length 8–11 mm.

Description: South and Central Americas. Small grey-brown moth, wings narrow, held tight furled to body when at rest. Forewings brownish or purplish grey, vaguely streaked with dirty cream marks. Hind wings uniform pale grey.

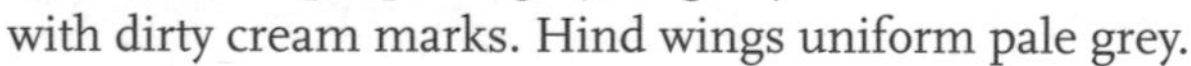

Life history: Lives in the fur of sloths (mainly three-toed, *Bradypus* species). Once a week sloth descends tree, scrapes small hollow with vestigial tail, defecates and covers dropping with leaf litter. Female moths fly down and lay eggs in the dung. Larvae feed in the dropping. Emerge some weeks later and fly up into canopy to find new sloth host.

Ecological notes: Not a commensal (minor non-damaging parasite) – sloth and moth are mutualists; both derive advantage. Moths get readily available food supply for their larvae. Sloths also benefit from moths, which bring biomass into the fur where green algae grow. This camouflages the slow-moving sloths in the canopy. They groom themselves and derive significant nutrients from eating the algal bloom in their fur.

Owl pellet moth (tapestry moth), *Trichophaga tapetzella*

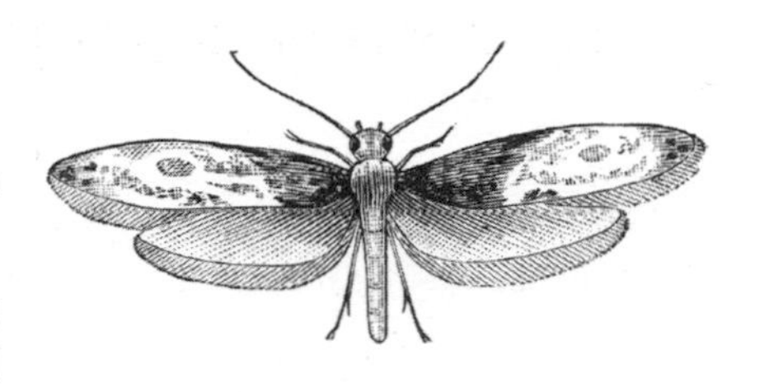

Size: Length 5–9 mm, wingspan 13–22 mm.

Description: Small grey and white moth; generally mottled, but basal third of wings darker, brownish or purplish grey. Wings held furled tight against body when at rest. Bird-dropping mimic.

Life history: Small pale caterpillar feeds in a sock-like case woven of its own silk, which it carries about as it crawls. Lives in bird and animal nests, feeding on moulted fur and feathers.

Ecological notes: Sometime minor household pest, eating woollen fabrics (including tapestries), rugs, horse hair furniture stuffing, pillow feathers. Regularly reared from owl pellets, where the larvae feed on the undigested feather and fur remains coughed up. Strictly speaking these are not dung, but droppings in a much broader sense.

DICTYOPTERA – TERMITES AND COCKROACHES

Termites, *Microtermes, Odontotermes, Macrotermes, Synacanthotermes* and others

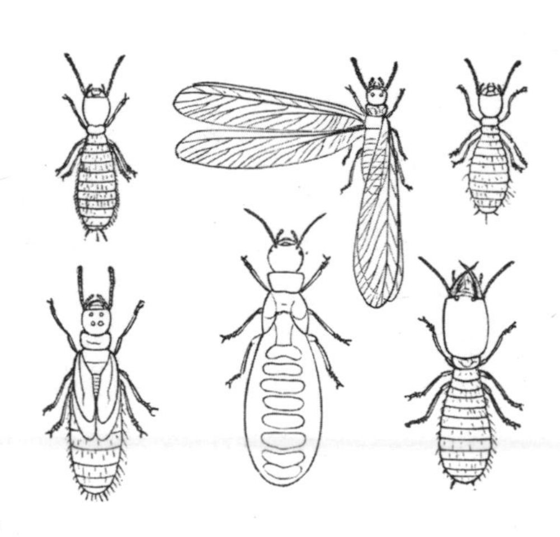

Size: Length 1–15 mm.

Description: Small pale creamy or yellowish-brown insects. Head relatively large on soft, narrow, short-legged body. Sometimes called white ants for their vague resemblance, large numbers and nest formation, but are unrelated.

Life history: Complex colony of fertile male and female and many thousands of sterile females (workers), of different shapes and sizes adapted to tasks of foraging, tending young, nest building, fighting. Build large nests of soil mixed with saliva to make tough concrete, elaborate structures to allow ventilation and food storage, collect plant material. Can digest cellulose because of gut micro-organisms. Some species culture fungus on stored leaf fragments in the nest and eat fruiting bodies.

Ecological notes: Only considered a dung-feeder since 1970s when they were discovered to be major recyclers of dung during African dry season when dung beetles less active. Removes remains of elephant, cow, camel and many other types of dung, treating the quickly drying droppings as merely part-processed plant particles.

Cave cockroaches, guanobies, *Trogoblatella*, *Spelaeoblatta*, *Eublaberus* and other species

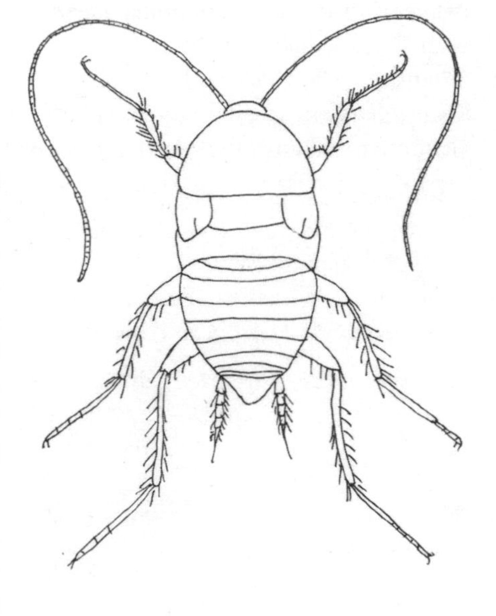

Size: Length up to 50 mm.

Description: Broad, flat, long oval insects, with long legs and very long, many-segmented antennae. Some species are winged, with leathery overlapping wing-cases, although these are sometimes shortened, and many flightless species lack them completely. Tip of abdomen with cerci – short antenna-like feelers.

Life history: General scavengers, eating whatever comes their way. Eggs laid in ootheca, a tough case made of hardened protein material secreted around the eggs inside the female abdomen.

Ecological notes: Cockroaches are a major component of cave faunas around the world, where bat droppings accumulate on the cave floors. They feed on the guano, on moulds, dead insects, dead bats and anything else. Some cockroaches have become domestic pests across the world. These started off as generalist detritivores in tropical leaf-litter rather than as cave species, but their sometime attraction to sewers and human latrines makes their kitchen-visiting habits unhygienic and unhealthy.

HYMENOPTERA – WASPS AND ANTS

Social wasp, yellowjacket, *Vespula vulgaris* and other species

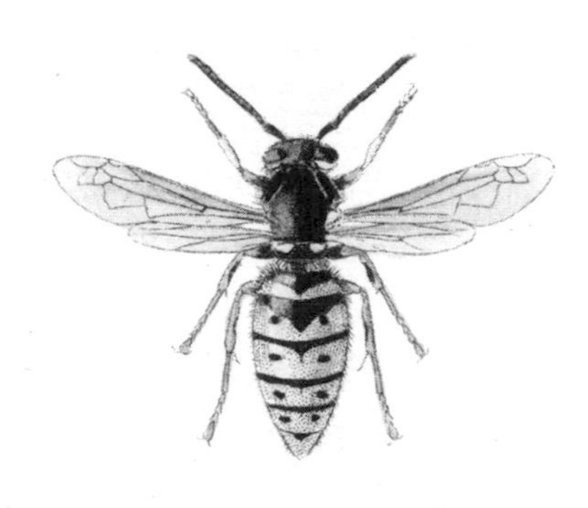

Size: Length 20 mm, wingspan 40 mm.

Description: Distinctive long, narrow, strikingly marked, black and yellow insects, with notable narrow 'waist' between thorax and abdomen. Head with large kidney-shaped eyes, stout antennae, yellow face and sharp jaws. Wings (large pair plus smaller pair) clear, membranous, folded into pleats when at rest.

Life history: Complex colony formation with single fertile female (queen) laying eggs and up to several thousands of sterile females (workers) foraging and nest building. Late in season new males and queens are produced which mate. Mostly predators of small insects, chewed bits of which are fed to the grubs in the nest combs, but some flower visiting.

Ecological notes: Not strictly dung feeders, but will visit carrion and other putrescent decay to feed. Sometimes sits in wait by a dropping to attack the dung flies and dung beetles that are attracted.

Scarab wasps, *Tiphia femorata, T. minuta* and others

Size: Length to 12 mm, wing length to 9 mm.

Description: Narrow, black, shining wasp-like, with narrow waist. Legs red or black.

Life history: Burrows into the soil (mostly sandy places) to find find dung beetle larvae or pupae, on which it lays its eggs. The maggots then devour the host grubs.

Ecological notes: Not strictly a dung-feeder, but definitely part of the dung community. Not generally associated with fresh dung, but a latecomer, after the dung has been partly or wholly removed.

Digger wasp, *Mellinus arvensis* (and other species)

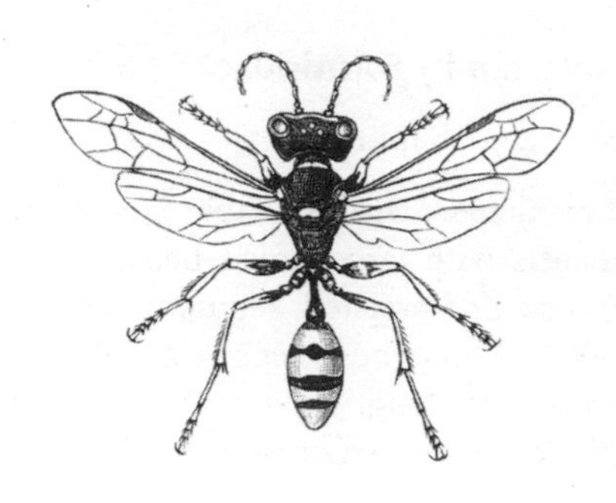

Size: Length 15 mm.

Description: Shining, narrow, black wasp, prettily marked with bright yellow flashes on head and thorax, and bars across slightly bulbous abdomen. Legs yellow. Wings, slightly brownish.

Life history: Captures and kills insect prey to stock a small burrow nest made usually in sandy ground. Egg laid in each prey-stocked cell, maggot eats store and emerges the following year. Each female works alone, but many gather in loose aggregations in suitable bare ground to form a 'wasp village'.

Ecological notes: Not a dung-feeder, but regularly observed sitting on fresh dung waiting to pounce on the blowflies, greenbottles and dung flies, often larger than itself, which are attracted. Mad scramble at the pounce, uses formidable jaws and sting to subdue its prey.

Parasitoid wasps, families Pteromalidae, Braconidae, Ichneumonidae, etc.

Size: Length 0.5–35 mm.

Description: Vast array of minute to large, slim, usually dark-bodied, sometimes metallic, wasp-like insects. Wings, one large pair, one smaller pair behind, usually clear membranous. Smaller species run-hop, larger species hawk low, quartering the ground looking for host victims.

Life history: Internal parasitoids of other insects. Eggs are laid direct into insect eggs, larvae or pupae and the hatching maggots eat the host alive, from the inside, finally killing it. Very often host-specific, each parasitoid species only attacking one particular host species, genus or close-knit guild of organisms.

Ecological notes: A ubiquitous, but poorly studied group of insects, despite many being large and brightly coloured. Virtually every insect in the world will have its own particular parasitoids. Many dung-feeding beetles, flies and other insects have their own parasitoid species attacking them, but rearing records are pitifully few.

Ants, family Formicidae

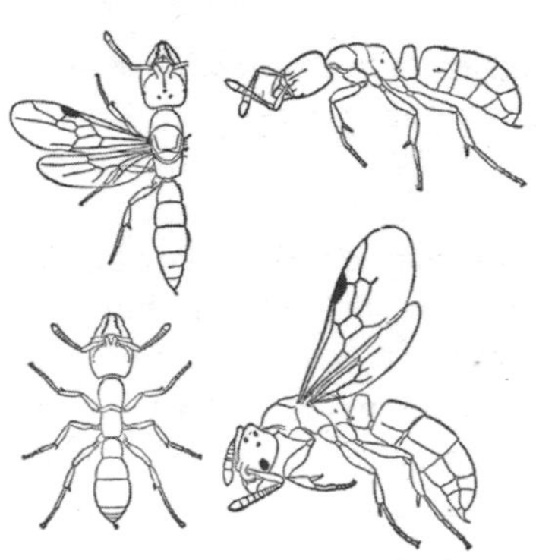

Size: 2.5–25 mm.

Description: Unmistakable ant-shaped insects with large broad head, narrow thorax and waisted segment (petiole), bulbous abdomen. Legs relatively long. Antennae long, elbowed in middle. Black, brown, red or yellow, or blotched combinations.

Life history: Complex colonies with fertile female (queen) laying eggs and large numbers, sometimes many thousands, of infertile females (workers), foraging, nest-building and brood-caring. Usually nest in soil, occasionally making small or large mounds. Most temperate ants feed on aphid honeydew (excess, barely processed plant sap, passing through the body), on plant leaves and stems, but also on roots.

Ecological notes: Not dung-feeders, but regularly found sheltering, or even nesting under old dung. A few predatory species may be after small dung organisms. May also be attracted by moisture in fresher droppings.

OTHER INVERTEBRATES

Shorebug, *Saldula orthochila*, and other species

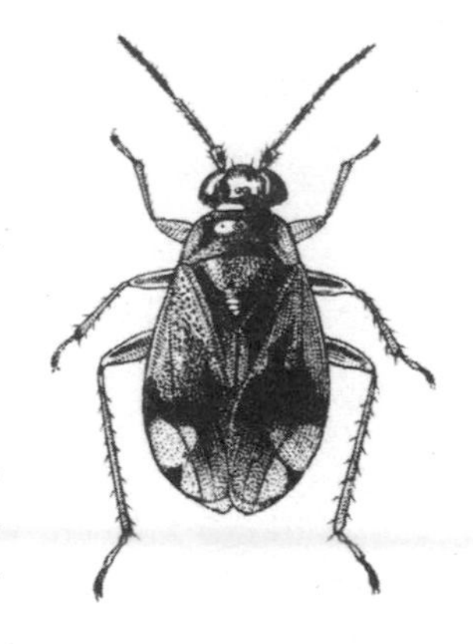

Size: Length 3–5 mm.

Description: Small, oval, flattish bug. Black or very dark brown, mottled with pale speckles. Prominent whitish spot on margin of forewing. Legs pale. Many very similar species.

Life history: Fast, active, hop-flying predatory bug. Most species hunt through the herbage, or across bare mud at pond and stream edges, or salt-marshes, hence common name.

Ecological notes: In northern Europe, this is about the only shorebug species known to inhabit dry places, such as wastelands, dunes and fields, and turns up surprisingly often under fairly fresh horse, sheep and cow dung, perhaps relieved by the moisture.

Lesser earwig, *Labia minor*

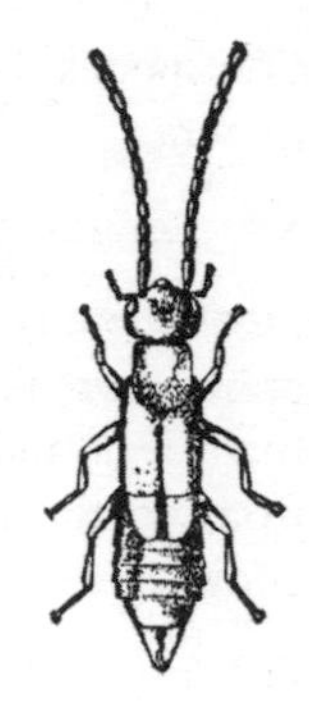

Size: Length 4–7 mm.

Description: Small, parallel-sided, vaguely flat insect. More or less uniform pale brown, head and antennae darker. Similar to the very characteristic and well-known common earwig (*Forficula auricularia*), but smaller, neater, more compact, less shining, with tail-tip forceps shorter, and less curved.

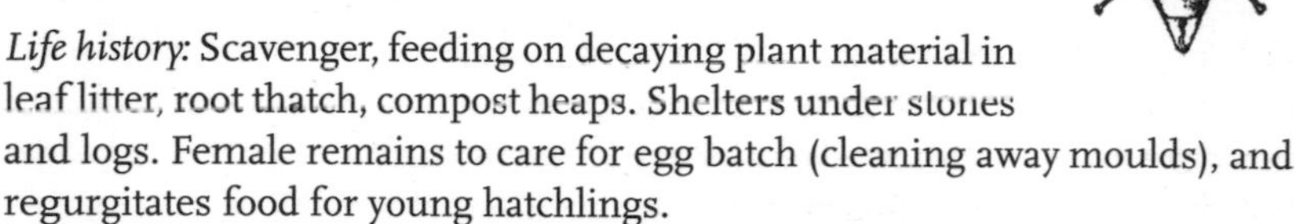

Life history: Scavenger, feeding on decaying plant material in leaf litter, root thatch, compost heaps. Shelters under stones and logs. Female remains to care for egg batch (cleaning away moulds), and regurgitates food for young hatchlings.

Ecological notes: Not a true dung-feeder, but commonly found in manure heaps. Once a very common urban species, in the days of horse transport, living in the frequent heaps of dung cleared from roads. Flies readily. *Forficula* is sometimes found sheltering under old dung.

Cave crickets, *Ceuthophilus, Caconemobius, Hadenoecus* and other genera

Size: Body length 15–20 mm.

Description: Characteristic cricket form, with short stout body and long legs, especially hind pair. Antennae very long. Cave species typically lack wings. Colour pale brown to near white.

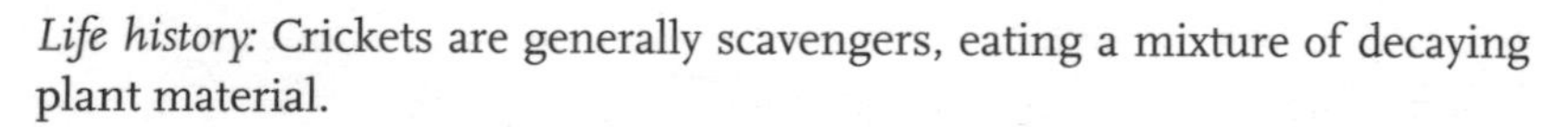

Life history: Crickets are generally scavengers, eating a mixture of decaying plant material.

Ecological notes: Many crickets have become adapted to cave life, feeding on bat guano, but also scavenging dead insects and whatever other organic matter they can find. In North America, bats are less common in caves, so cave cricket frass has become a specialist microhabitat for yet another layer of organisms. Surface species 'sing' by rubbing wings together in courtship displays, but flightless subterranean species have lost this ability. Many have also lost ability to hop, and although retaining long hind legs, the musculature appears to be atrophied.

Centipedes, *Lithobius*, *Haplophilus* and other genera

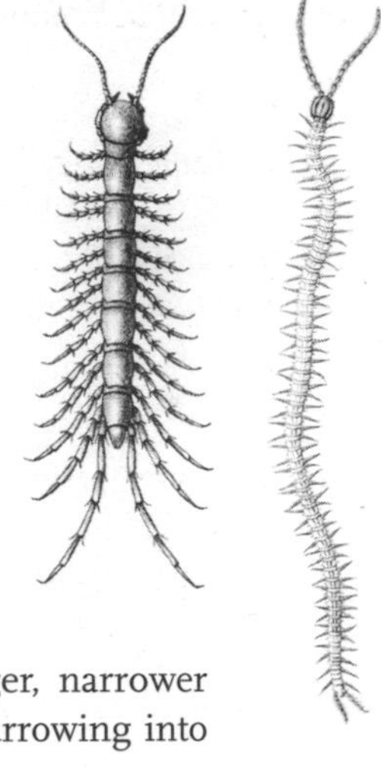

Size: Length 18–80 mm.

Description: Long, narrow, usually pale yellowish, orange or pinkish-brown worm-like animals with many legs, and long antennae. Leg numbers vary from 30 to 200, but only one pair of legs per body segment.

Life history: Predators after small invertebrates in the soil, leaf-litter and root thatch. Shorter, broader species with fewer but longer legs run fast over the ground. Sharp jaws are modified limbs reaching around the head capsule. Longer, narrower species with more but shorter legs adapted to burrowing into the soil or litter layer.

Ecological notes: Regularly invade old dung after prey, or use it as shelter.

Millipedes, *Polydesmus*, *Tachypodoiulus* and other genera

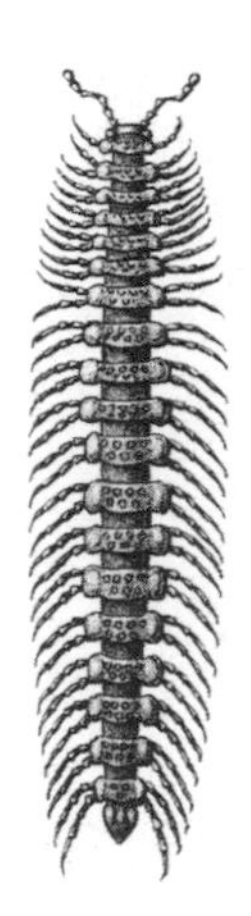

Size: 15–60 mm.

Description: Snake millipedes (*Tachypodoiulus*, etc.) cylindrical, shining, black (sometimes marked with red or yellow patches), up to 250 legs. Flat-backed millipedes (*Polydesmus* etc) gnarled, knobbled, appearing flat because of flange-like edges, about 80 legs. All millipedes have two pairs of legs per body segment.

Life history: Feed on decaying plant material in leaf-litter, root thatch, under logs, etc.

Ecological notes: Sometimes found under old dung and probably genuinely eating the decaying remains of the plant material within it. In a curious twist of ecological fate, some millipedes produce a secretion to which a few species of dung beetles are attracted. Some beetle species have evolved to become specialist dead millipede scavengers, and one, *Deltocheilum valgum*, is now an obligate millipede predator.

Mites, Order Acari

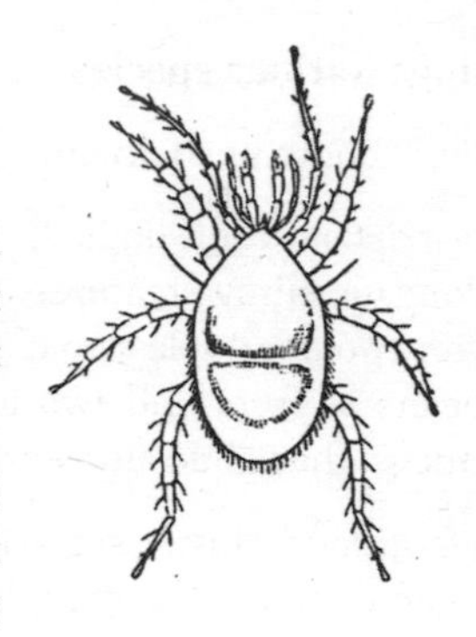

Size: Length 0.1–5.0 mm.

Description: Microscopic to minute, round or broad oval, dull yellowish through brown to almost jet black. Body seemingly only one segment. Eight legs.

Life history: Huge diversity but those encountered in dung are either feeding on the decaying plant material, or sucking the body fluids of other dung inhabitants.

Ecological notes: Tiny (0.5 mm) shining black convex oribatid mites are very common in the soil and leaf layer or under rotten logs, and feed on decaying plant material. Large dung beetles are often infested with many larger (1–2 mm) pale mites, which use the adult beetles as transport onwards to the next pat. Sometimes these attached mites are sucking body fluids through chinks in the beetles' armour, although many disperse on landing to attack much softer fly larvae.

Woodlice, *Oniscus asellus, Porcellio scaber* and other species

Size: Length 2–20 mm.

Description: Familiar broad, domed, multi-segmented creatures, with relatively long antennae and 14 short legs. Mostly grey, but some flecked with lighter, brighter colours including pink, yellow or orange.

Life history: Scavengers feeding on rotten plant material. Common in leaf-litter, under rocks and stones, logs and loose bark of dead trees.

Ecological notes: Often found under old dung, and probably feeding on the rotting detritus. They may just be sheltering in a dark, damp place, since they are just about our only terrestrial crustaceans and liable to desiccation in hot weather.

Slugs, various species

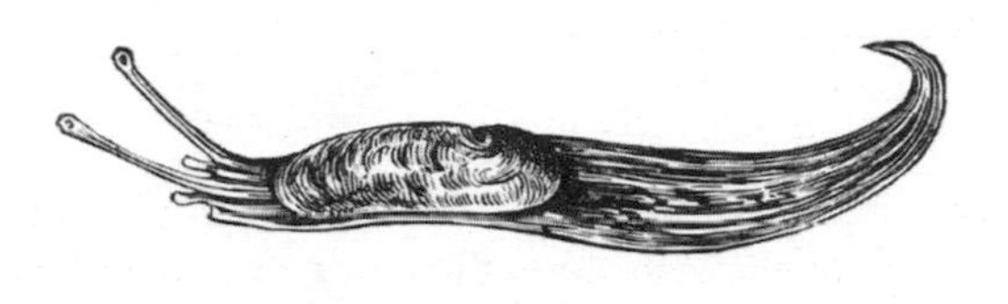

Size: Length up to 20 cm.

Description: Soft-bodied, elongate, slimy creatures with two short telescopic feelers at front and two longer telescopic eye stalks above. Covered with mucus, they glide along on silvery lubricated trails.

Life history: Mostly scavengers in soil and leaf-litter, eating any decaying organic matter, but many are predators on other slugs, and snails. Lacking the protective but bulky shell possessed by snails, slugs can squeeze into the tightest of spaces and are mainly subterranean, burrowing into the soil, but emerging to feed at night or in moist weather.

Ecological notes: Often found under old dung and may be feeding or sheltering, but also attracted to fresh dung, especially that of carnivores such as cat, fox and badger.

Nematode worms, various species

Size: Length 0.1–2.0 mm.

Description: Microscopic to minute, apparently unsegmented hair-like worms, circular in cross-section. White, black, brown or other colour.

Life history: Mostly parasitic inside a vast diversity of other animal species, including insects.

Ecological notes: Those species found in dung are usually gut or other parasites in the animals that dropped the dung. Often these are the larvae (or eggs), which only develop further if they are reingested by new hosts eating grass contaminated by dung, or on which the infectious stage has lodged after the dropping has been weathered, removed or recycled. Sometimes there is an intermediate host, for example *Gongylonema* eggs are eaten by dung beetles, which must then get accidentally eaten by cattle, where they infest the digestive tract.

Earthworms, *Lumbricus terrestris* (lobworm) and other species

Size: Length to 30 cm.

Description: Large, soft, many-segmented worm, bruised pinkish with purple or blue hints. Cylindrical, with swollen 'saddle' (containing the reproductive organs) about one-quarter from front end. Tail blunt and flattened at tip.

Life history: Creates temporary mucus-lined burrow in the soil, from which it half ventures out onto the surface at night to feed. It grips leaf fragments in its mouth, and draws them down into the burrow to feed. Two hermaphrodite worms mate, above ground, at night, by aligning saddles and exchanging sperm.

Ecological notes: Soil-dweller (sometimes called the night crawler) and plant-feeder, but often found sheltering under dung, on particles of which it will feed. In temperate zones worms are late arrivers at the pat, but major consumers of the leftovers.

Brandling or tiger worms, *Eisenia fetida*

Size: Length to 20 cm.

Description: Long, narrow, multi-segmented worm, bright red, often appearing banded red and pink, or orange. Slightly swollen saddle area much less pronounced than in lobworm.

Life history: Not a soil-burrowing worm, but a denizen of the leaf-litter layer. Eats decaying plant material.

Ecological notes: Not strictly a dung-feeder, but often found in manure heaps, as well as compost bins and piles of grass-cuttings. This is the worm which is commercially available for wormeries, free-standing layered composting systems for the garden.

OTHER ANIMALS

Dog, *Canis familiaris*

Size: Hugely variable from toys 6 cm at the shoulder to giant mastiffs over 1 m high.

Description: Varying from stunted squat to slim and muscular or nearly skeletal. Colours any combination of black, brown and white. Short or shaggy fur. Pert or flop-eared.

Life history: Familiar family pet, domesticated from grey wolf, *Canis lupus*. Still loosely carnivorous and behaviour controlled by pack mentality in the human pack home.

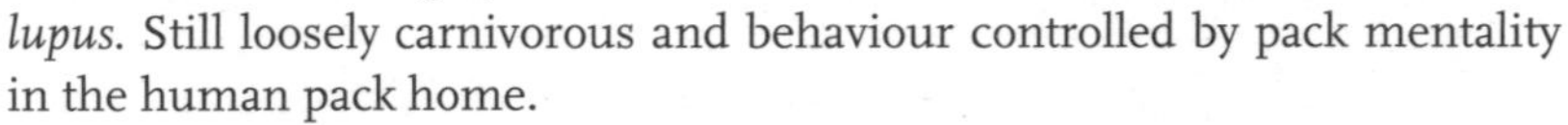

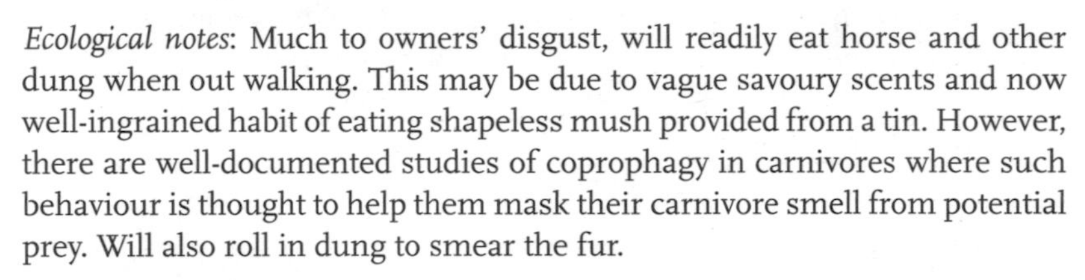

Ecological notes: Much to owners' disgust, will readily eat horse and other dung when out walking. This may be due to vague savoury scents and now well-ingrained habit of eating shapeless mush provided from a tin. However, there are well-documented studies of coprophagy in carnivores where such behaviour is thought to help them mask their carnivore smell from potential prey. Will also roll in dung to smear the fur.

Badger, *Meles meles*

Size: To 90 cm long, 30 cm high, weighing 15–17 kg.

Description: Stocky, short-legged, broad-shouldered but narrow-snouted mustelid, mostly grey blend of black and white hairs, but with distinctive black and white striped head.

Life history: Large deep burrow nest in the soil all year. Young born in almost any month, though spring peak. Nocturnal foraging after fruit, roots, insects, other small animals, wasp and bee nests.

Ecological notes: Not a dung-feeder, but taste for earthworms means they turn over or dig into old dry cow pats, especially in damp meadows, taking any insect grubs they find too.

Rook, *Corvus frugilegus*

Size: Length 40–47 cm, wingspan to 90 cm.

Description: Large, handsome, black bird with bluish tinge to feathers. Face bare, grey. Beak narrow, grey. Leg bases shaggily feathered giving impression of wearing ragamuffin trousers.

Life history: Nests high in trees, usually in large groups (rookeries). 3–9 eggs laid in March or April, young fledged in 5–6 weeks. Scavenges for whatever it can find – insects, worms, carrion, fruit, seeds.

Ecological notes: Often seen pecking at seasoned cow pats, where it appears to search for larvae of the noon fly, *Mesembrina meridiana*. Probably eats other dung-dwellers it comes across. Crows, jackdaws, choughs and magpies have similar behaviour.

Egyptian vulture, *Neophron percnopterus*

Size: Length to 65 cm, wingspan to 165 cm, weight to 2.8 kg.

Description: South-west Europe, North Africa, Arabian Peninsula, India. Large pale grey, or dusty brown, to bright white bird with black flight feathers in the wings. Unfeathered face bright yellow.

Life history: Scavengers, eating flesh from animal carcasses, also vegetable material, insects and dung.

Ecological notes: Often seen pecking at cow or goat dung, earning it the local name *churretero* or *moniguero*, meaning 'dung-eater' in Spain. The birds obtain carotinoids, yellow pigments, from the faeces which give their faces the bright colour.

Sitatunga, or marshbuck,
Tragelephus spekei

Size: Length 115–170 cm, height 75–125 cm at the shoulder.

Description: Rather shaggy, slim-faced antelope; chestnut red to brown, males with rough main and pale dorsal stripe. Face with white bar below eyes. Males with prominent spiral horns. Hooves splayed.

Life history: Central African. Secretive, in the dense vegetation of the marshy waterways of the Congo, Cameroon, etc., adapted to swamp dwelling by waterproof shaggy coat and splayed hooves which allow them to walk well on mud and floating islands of reed, papyrus, and other water plants.

Ecological notes: Although, like other antelopes, they browse leaves, stems and shoots, sitatunga spend a significant part of their time foraging in elephant dung, eating the barely digested leaf material and also seeds which pass through the elephant digestive tract.

Alabama cave fish,
Speoplatyrhinus poulsoni

Size: up to 6 cm.

Description: Slim, unpigmented fish, appearing translucent or slightly pinkish. Blind, lacking eyes. Head blunt, snout flattened. Lacks branching fin rays, fin membranes deeply cut in.

Life history: Occurs only in Key Cave, Lauderdale, Alabama. Poorly known. First discovered in 1974. Population may be less than 100.

Ecological notes: May feed on guano dropped by grey bat, *Myotis grisescens,* which represents the only biomatter entering the cave. Also likely to eat the other cave-dwelling animals which rely on the nutrient-rich guano, mostly crayfish, isopods and copepods.

Grotto salamander, ***Eurycea spelaea***

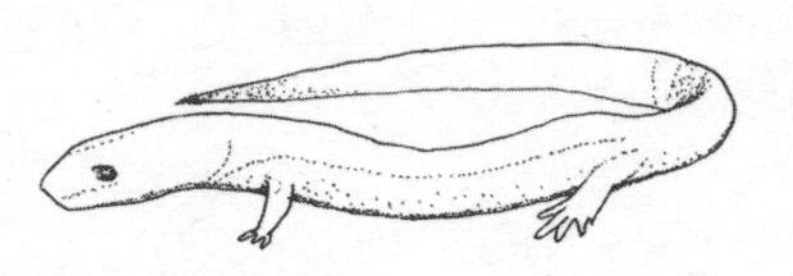

Size: Length to 13.5 cm.

Description: Adult pinkish white, but larvae brown or purple, often flecked with yellow. High tail fin, external gills.

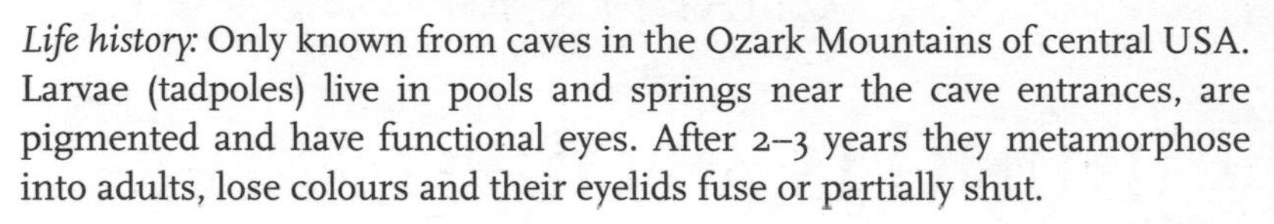

Life history: Only known from caves in the Ozark Mountains of central USA. Larvae (tadpoles) live in pools and springs near the cave entrances, are pigmented and have functional eyes. After 2–3 years they metamorphose into adults, lose colours and their eyelids fuse or partially shut.

Ecological notes: Although they probably feed on whatever else they can find in the cave pools, including invertebrates and scavenged material washed in during storm floods, they have often been observed feeding directly on bat guano, and have been measured obtaining nutrition from the droppings. Other amphibians have been observed eating guano, but this was thought to be merely to derive gut bacteria and other micro-organisms.

CHAPTER 13

DUNG IS A FOUR-LETTER WORD – A SCATOLOGICAL DICTIONARY

Billets, or **billetings**, etymology highly suspect, but used in old books to mean the dung of the fox. Probably only ever used by medieval hunters.

Bullshit, pretty obvious blunt derivation. Slightly vulgar term meaning nonsense, but implying a deliberate attempt to mislead. See 'chickenshit' and 'poppycock'.

Cack, from the Latin *cacare*, and Greek κακκη (*kakkh*), to void excrement, and the excrement itself. Similar to the infantile-sounding 'caca' (or 'cacca') used by Italian children. Related to the Greek κακος (*kakos*), meaning bad or evil, from which we get words like cacophonous. This puts another slant on the notion of cack-handed meaning more than being just a bit awkward. A caccagogue is an ointment of alum and honey, used as a laxative. *Caccobius* is a genus of dung beetle (Scarabaeidae), and *Hypocaccus* a genus of clown beetle (Histeridae).

Caccobius, lives under cack.

Caecotrophy, from the Latin *caecum* originally meaning blind, and Greek τροφιψος *trophicos*, pertaining to food and nutrition. In anatomy the 'caecum' is the 'blind' gut, a pouch at the beginning of the large intestine, reduced to the appendix in humans, but a large digestive organ in many herbivores. Caecotrophy is the reingestion of special mucus-covered faeces (caecotropes), especially by rabbits, to obtain additional nutrients on a second pass through the gut.

Call of nature, idiomatic expression of a general need to urinate or defecate. The only euphemism I will allow myself, but only if accompanied by the natural act itself, somewhere out in nature. See '*stercore humano*' below.

Cast, from Middle English *casten*, to throw, or the thing that is thrown. Worm casts are the digested remains of the worm's meal, quite literally thrown up onto the surface of the soil by the animal at the entrance to its burrow, into a neat twisted tubular tangle. Arguably this is what soil is, at least the humus layer that covers most of the vegetated world. Charles Darwin spent many a happy year measuring exactly how much soil was cast up by earthworms.

Castings, similar to 'cast', but in this case the pellets thrown up by owls, falcons and other birds of prey. Technically not dung (although similarly shaped) as they are ejected from the oral, rather than the cloacal end of the bird, and comprise undigested skin, fur, bones, gristle, sinew and feathers.

Cesspit, or **cesspool**, origin uncertain, but maybe a corruption from 'sus-pool', a place where hogs wash, *sus* being Latin for pig. A pit, or covered tank for receiving domestic sewage, allowing solids to be biodegraded by bacterial action and water to seep out into the surrounding soil.

Chickenshit, similarly straightforward derivation to 'bullshit', but rather than implying deceit, it suggests something so trivial as to be inconsequential. See 'poppycock'.

Commode, from the Latin *commodus*, 'convenient'. A chair or other piece of furniture containing or concealing a chamber pot. Nowadays only used in hospitals, or where some incapacitated person needs the immediate convenience of a discreetly disguised toilet close by.

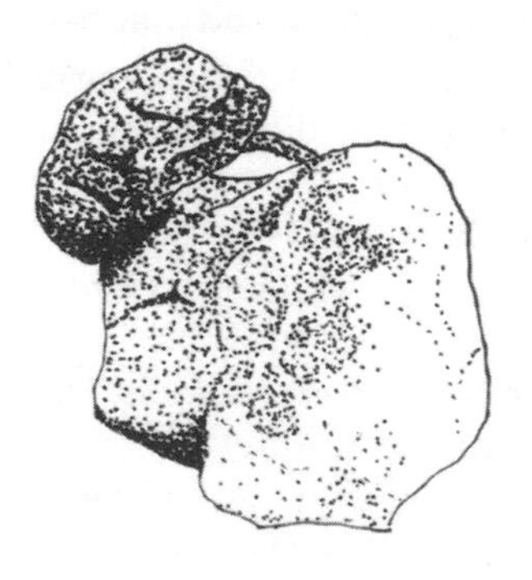

Sadly, when researching this book I was unable to find a specific technical term for chicken droppings.

Coprology, from Greek κοπρος (*kopros*), dung, and λογος (*logos*), a discourse. The study of dung, although some less erudite dictionaries claim it means the use of obscenity in speech, literature or art. Similar are 'coprophagous' (eating dung), 'coprovore' (a dung-eater), 'coprophilous' (dung-loving), and 'coprolite' (fossilised dung). *Copris lunaris* is the handsome English scarab, now sadly extinct in England. *Coprinus* is a genus of mushroom, the ink caps, often found growing out of old pats, but delicious to eat.

Crap, from Middle English and Dutch *krappe*, originally rubbish, bits cut off, sifted or thrown away, and at one time mainly used to mean chaff of various cereals, and the residue left by brewing beer, or the dregs in the barrel (see 'faeces'). It's tempting to link it to 'scrap', in the sense of leftovers, discarded or rejected rubbish. Now vulgar slang for excrement, and the act of dropping it, though 'crappy' still retains the original meaning of not very good. In Britain, at least, slightly less abruptly expletive than 'shit'. Nothing to do with Thomas Crapper, sometime claimant to the invention of the water closet.

Crottels, also **croteys**, **crotisings**, etc., seemingly from the Old French *crotte*, the larger pelleted droppings of sheep and goat, but taking the diminutive ending for the smaller round pellets of rabbit and hare. Several alternative archaic spellings include 'croteys' and 'crotisings'. Mostly in ancient books on hunting or rabbit husbandry, but a word I'd like to see repopularised. Several French cheeses take the name *crottin*, which even though this translates as 'dung' appear to be quite popular. Their roundish, blob-shaped forms resemble goat droppings. Maybe they're ironic.

Dags, from Middle English, something that hangs, related to 'tags'. The small pellets of dung caught in the matted wool under the tail of an unkempt sheep. Mainly Australia and New Zealand.

Defecate, the act of dropping or passing faeces (see below). Mostly in formal prose in medical, zoological or scatological textbooks and other scientific literature.

Diarrhoea, from Greek δια *dia* 'through' and ρεω *rheo* 'a flow'. Quite literally a flowing through. Sometimes called 'flux' by literary types. No more needs to be said.

Dirt, from Norse *drit*, old Dutch *driet* and modern Dutch *drijten*, to void excrement. Almost entirely with changed meaning nowadays, and used for the earth layer in which plants grow, or the stains made from it. About the only time it ever means dung now is the occasional coy or euphemistic tabloid exclamation about dog dirt. A similar transformation has occurred with 'soil', see below.

Doll, of Scottish origin, therefore of doubtful etymology, meaning dung, particularly that of pigeons.

Dreck, also **dregs**, possibly from Norse *dregg* or Swedish *drägs*, meaning to draw, and implying leftovers, or sediment after liquor is drawn off. Still familiar to tea and beer drinkers who might leave dregs at the bottom of cup or barrel. Now rather obsolete for dung.

Dropping, something that is dropped, from the Old English *dropian* or *droppian*. Although potentially blighted by the idea that this might be a euphemism (and therefore pathetically prudish), farmers and hunters have long referred to droppings in a purely down-to-earth fashion. The *OED* suggests it should now only be plural, but entomologists, at least, commonly refer to an individual, rather than to a scattering of them. So that's all right.

Dung, self-explanatory if you've read this far in the book, and my descriptor of choice, from the old English *dung* or *dyngian*. See 'midden'. Not an expletive in any sense. The *OED* suggests hyphenating dung beetle and dung fly, but until recently entomologists have kept words separate if they are what they say (so dung flies truly are flies), but joining those that are not (butterflies are not flies). This, however, is much misunderstood, often flouted, and probably not really very historically robust.

Easement, from the Old French *aisement*, after *aisier* to ease, and literally meaning anything that gives relief. If it were not so delightfully archaic it could easily be dismissed as an unnecessary euphemism. Ought to be more popular. Look out for it in gritty historical novels.

Earth closet, a fairly basic toilet, just a step up from a latrine, where earth is used to cover the excrement deposited in the hole. See 'water closet'.

Egesta, from Latin *egestus* from *egerere*, to expel. Waste matter ejected by the body. Very obscure, archaic, now very rare, even in medical texts.

Excrement, from the Old French *excrément* or Latin *excrementum*, and meaning, in very physiological terms, what remains after the body has sifted through food, and has subsequently been ejected from within. Often used in a cold biological or medical sense similar to excretion, from which we also get the very formal sounding 'excreta'.

Fewmets (also **fumets, fumes, fumeshings)**, from Anglo-French *fumets* or *fumez*, and the Latin *fimare* (dung), and used in some very old books on hunting to describe the droppings of the various quarry, but especially of deer. Not to be confused with 'fumet', a rich stock, usually of game or fish, used in cooking.

Faeces (or, in the USA, **feces)**, from the Latin *faeces*, the plural of *faex*, sediment or dregs, and related to 'faecula' or 'fecula' which, at the end of the 19th century, still meant a crust of wine, sediment or lees. Like 'excrement', its use is often in a hard, cold biological or forensic sense.

Fiants, **friants**, **fyants** or **fuants**, from the Old French *fient* and possibly the Latin *fimus* (dung). A delightfully archaic term, from old books on hunting, to mean the dung of the fox, or sometimes the badger. We could do with bringing this one back into fashion.

Fime, from Latin *fimus*, dung. Now wholly obsolete.

Fimicolous, from Latin *fimus* dung and *colere* to inhabit, growing in dung. Originally coined by botanists to describe plants growing in dung, but easily appropriated by entomologists. Also 'fimetarious'. The attractive black and red UK dung beetle *Aphidius fimetarius* is aptly named.

Frass, from the German *frass*, itself from *fressen*, to devour (now only used for animals), and specifically used for the droppings of insect larvae (particularly caterpillars) and the powdery detritus left by wood-boring insects. One of the many arcane terms an entomologist can throw into the conversation to show off.

Garderobe, from Middle English and Old French, *garder* to watch and *robe* clothing, and similar to toilet (see below) in originally meaning a dressing room (a wardrobe). Nowadays most familiar to visitors at ancient castles and manorial ruins where all that remains is a seat above a hole which would have taken ordure directly down into the moat.

Gong, **gonge** or **goonge**, from Anglo Saxon *gang*, a going, a passage, a privy. Wholly outdated and obsolete now, but occasionally used in the hyper-archaic term 'gong-farmer', someone who cleared and cleaned out cess pits or privies.

Guano, via the Spanish *guano* and *huano* from *wanu* or *huanu*, meaning 'dung' in the original Quechua language of the South American indigenous peoples. The long-term (centuries or millennia) accumulations of bird droppings, specifically the guanay cormorant, *Phalacrocorax bougainvillii*, along the western edge of South America. This extremely dry zone retains the excrement without it being leached by rain or removed by recycling scavengers. Its high phosphorus and nitrogen content made it valuable as a soil fertiliser. Also now used for similar long-lasting heaps of bat dung in caves.

Honeydew, combination of honey (which it tastes like) and dew (which it looks like). Clear or pale tawny liquid excreted by aphids (greenfly). In sucking so much plant sap to get the meagre protein content, aphid excrement is little changed from the watery juice in leaves and stems. Ants 'milk' the aphids in exchange for protection services. Some bumblebees and butterflies lap up the spilled honeydew where it drips onto the leaves.

Jakes, a privy, or outdoor toilet, etymology uncertain. Well used by Shakespeare, who may have punned the French name Jaques (deliberately mispronounced 'jay-quees') with it. 'Jacksie', very informal British expression for bottom, supposedly related, but since 'jack' is reputed to be the English word with the highest number of different meanings, this may be wholly coincidental.

Latrine, from the French and Latin *latrina*, itself a contraction of *lavatrina*, originally from *lavare*, to wash (see 'lavatory'). Traditionally a temporary hole or trench dug in the ground for the sole purpose of receiving excrement, usually a communal facility at campsite or military base. The collection of small shallow dung pits dug by badgers is aptly described as a latrine.

Lavatory, Middle English, from Latin *lavatorium*, a place for washing (from *lavare*, to wash). Originally a wash house, with baths and/or laundry, but a toilet since the 19th century. Sometimes shortened to 'lavy', 'lavvy' or 'lavvie'. I grew up thinking that the handsome pink-flowered tree mallows, *Lavatera* species, were named because, in times past, they had been planted near latrines, privies and lavatories, so that the soft leaves could be used instead of toilet paper. I now discover that I have been cruelly misled, and in fact they are so called after the Laveter brothers, renowned 17th-century Swiss physicians and naturalists.

Laystall, also **laye-stowe**, **ley-stall** and **loi-stal**, a joining of stall (in the sense of a stable) and what lays on the floor inside it – dung, or a dung heap. Obscurely dialectical (hence variant spellings). See 'stallage', below.

Lesses, from French *laisées*, leavings, and related to 'lees', the dregs in a wine bottle. The dung of wild boar.

Loo, origin unknown, but first appeared about 1940. Possibly a contraction of 'Waterloo', a punning trade name for a wily cistern and water closet manufacturer. A thoroughly British word, pompously coy and socially egalitarian in equal measure.

Make water, a tired euphemism for 'urination'. Now sounding like a line from *Carry on Nurse* or some such 1950s blather.

Manure, from Anglo-French *mainoverer* and Old French *manouvrer* (related to 'manoeuvre') meaning to cultivate land, and mostly used as verb, or noun to mean composted animal dung, mixed with hay, straw or sawdust and used to fertilize the soil.

Merde, French expletive, from the Latin *merda*, dung, more or less equivalent to, and meaning, 'shit'. Since it's in a foreign language, it can be safely used in a knowing humorous way, as less vulgarly offensive than either 'shit' or 'crap'. The only even vaguely associated word in English appears to be the obscure 'merdigerous', meaning carrying dung, in the way that tortoise beetle and lily beetle larvae cover themselves in their own excrement. I read in a book, so it must be true, that in Alsace, on the German–French border, horse dung was known by the mixed language term 'Pferde merde'. Several dung beetles take the specific epithet *merdarius*.

Micturation, from Latin *micturire*, to urinate. Rather obscure nowadays, and generally limited to technical, medical or forensic texts.

Midden (also **mixhill**, **mixen**, **myxen** and **myxene**), Middle English *myddyng*, from Scandinavian *myk-dyngja*, literally a muck heap. Essentially a general organic refuse heap, kitchen, farmyard, faecal, or all three. Nowadays mostly used in the historical or archaeological sense of excavating ancient rubbish dumps to discover how our predecessors lived, and what they ate.

Muck, Middle English *muk*, possibly from similar origins as 'midden' from Scandinavian *myk-dyngia*, supposedly related to Old Norse *myki* meaning dung, itself from Old German root *muks* meaning 'soft', hence 'meek'. My mother is the daughter of a second-generation north Kent farmer and when I started writing this book I approached her to see if they had used any specialist terms for the dung of particular animals, or any local dialect words. Her response: 'We just called it muck.' Yeah, thanks Mum.

Mutes, via French *mutir* and Old French *esmeutir* and *esmeltir*, Old Dutch *smelten*, *smilten*, to smelt, to urinate, and related to smelt in the sense of releasing (melting) liquid metal from ore. The excrement of falcons or hawks, although historically used for the droppings of many different birds. Mostly plural in modern bird of prey books, but singular used to be acceptable.

Night soil, see 'soil' below. Human sewage, supposedly collected discreetly under the cover of darkness so as not to offend the sensibilities. Either disposed of at the midden or dung heap, or used for manure, or in the leather tanning industry.

Number twos, infantile euphemism for defecation, along with 'plop-plops' and 'big jobs', and contrasting with urinary 'number ones'. My father told of his time as a very young child hospitalised with diphtheria in the early 1930s. Each morning the nurses would work their way along the ward, asking each bed-ridden child in turn whether it would be number ones or number twos that day. Childish repetition gives 'two-twos', which might have been corrupted to 'doo-doos'. In my own early schooldays I remember playing Lego with other 5-year-olds, and I could never understand why picking up a piece with two studs (obviously called a 'two') was fine, but playing with two 'twos' reduced several of my classmates to uncontrollable giggles.

Ordure, from old French *ord*, meaning 'foul', a corruption from Latin *horridus*, 'horrid'. Excrement in its widest sense, and also used for anything unpleasant or noxious.

Pat, Middle English *patte*, something flat. The flat, round dung of cattle. Also sometimes 'pie', 'pad' and (especially in North America) 'chips'.

Pee, childish or prudish contraction to the initial letter of the slightly more vulgar 'piss'. Similar to 'wee-wee'.

Po, from the French *pot de chambre* (chamber pot), a potty. A pot, usually porcelain or enamelled metal, for use in the bedroom, rather than having to go outside to the privy.

Poo (or sometimes **pooh**), infantile exclamation of disgust, now attached to faeces (sometimes as 'poop') and the act of defecation. Should only be used if talking to small children or tabloid journalists.

Poppycock, nonsense, silliness or downright twaddle. From the Dutch *pappekak* – dung (*kak*) which is soft (*pap*), or has come from a doll (*pop*).

Has the same connotations, if more gently expressed, as 'chickenshit' or 'bullshit'.

Privy, from the French *privé*, meaning private. A toilet, usually in a small shed, outside the house, in the garden. Originally anything private, secret, hidden or shared only between close associates, hence 'being privy to' some particular knowledge, or the UK monarch's Privy Council.

Public convenience, a public toilet, usually one built through municipal munificence for the benefit and convenience of the passing public. See 'spend a penny', below.

Restroom, North American euphemistic term for toilet, particularly one in a public building such as an office, hotel, restaurant or service station. Before I visited the USA, I always had vague imaginings that in such a wealthy country these must be luxurious places, sumptuously fitted, and with comfy sofas and chairs in which the resting might take place. Needless to say, I was later disappointed.

Road apples, North American slang term for horse droppings, alluding to the shape and size of the nearly round boluses of dung which scatter when dropped onto a hard road surface.

Rypophagous, from Greek ρυπος (*rypos*) dirt and φαγειν (*phagein*) feeding. Eating or subsisting on filth. Not an everyday word, and one I suspect has not been seriously used since the 19th century, when dung beetle study was on the ascendant.

Scarn, possibly also **sharn**, from Anglo Saxon *scearn*, and similar to Norse and Danish *skarn*, dung. According to old dictionaries it was obscurely dialectical, probably mostly Scottish, at the end of the 19th century. One online dictionary claims scarn-bee as a dung beetle. Who knows?

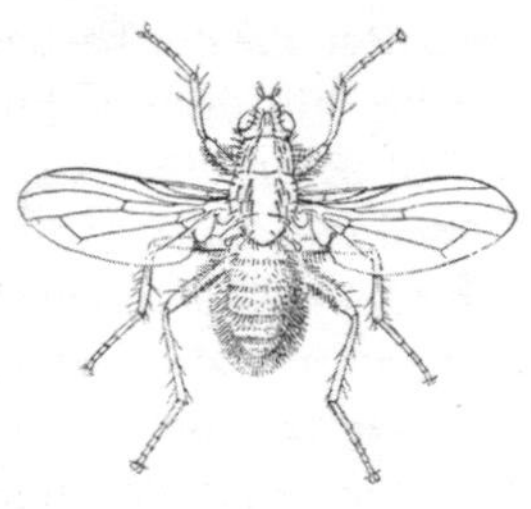

The ubiquitous yellow dung fly, *Scathophaga stercoraria.*

Scat, from Greek σκατ (*scat*), animal dung, especially that from a carnivorous mammal, such as otter, bear, wolf, fox. Mostly used by hunters and naturalists. This same root gives us 'scatology', the study of dung, and is immortalised in the name of the common yellow dung fly *Scathophaga* (dung-eating) *stercoraria* (dung-inhabiter). Skatole is one of the more fragrant chemical substances in dung giving it its distinctive smell.

Scumber, contraction of 'discumber', or 'disencumber', in the sense of getting rid of an encumbrance. Archaic term for voiding dung, also the dung itself, especially that of a fox.

Scybala, from Greek σκυβαλον (*skubalon*), dung. A hardened mass of faeces, now only used in the most technical of obscure medical books to describe the sort of thing that happens in extreme constipation. The scarce British dung beetle *Aphodius scybalarius* is now, sadly, known as *A. foetidus*.

Sewage (or **sewerage**). Combined effluent from domestic houses and public buildings, taken away by some sort of drainage system (see 'sewer' below), and including faeces, urine, flush water and general washing water; often incorporating rain run-off from gutters, road surfaces and other forms of hard standing. 'Saur', which may be related, is an archaic and now obsolete word for dirty water.

Sewer, via medieval Latin *seware*, and the Roman Latin *exaquare* from *ex* 'out' and *aqua* 'water'. Drainage system by which sewage is taken away. In the developed world most sewers are underground pipes, but traditionally they would have been river courses, then specially constructed ditches. Water-beetling, as a boy, on the Lewes Levels in Sussex, I was more than a little perturbed to discover that one of the big ditches I was exploring with a water net was called Celery Sewer. It took some time to convince myself that Lewes's effluent no longer passed through it.

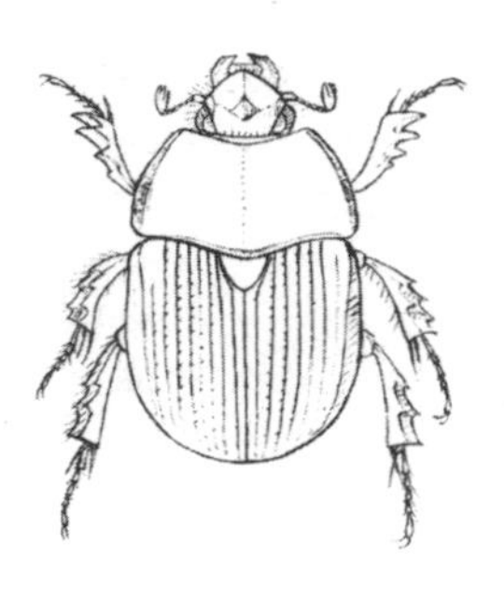

Dor beetle, *Geotrupes* species, shard-borne rather than shard-born.

Shard, Shakespearean for dung? Odd one this. Shakespeare uses the expression shard-born to describe dor beetles 'the shard-borne beetle with his drowsy hums (*Macbeth*, III.ii.42), and some dictionaries claim he meant born in dung. But it seems more likely to me that he meant shard-borne, i.e. carried on shards, meaning the stiff wing-cases which are like broken pot shards. Likewise in *Antony and Cleopatra* (III.ii.20) 'They are his shards, and he their beetle' implies a supportive, rather than an excremental relationship.

Shit, old English *scitte* 'diarrhoea', and similar to Dutch *schijten* and German *scheissen*. Excrement, and the act of depositing it. Now more often used as a vulgar expletive, possibly slightly more robust that 'crap'. 'Shitty', like 'crappy', is often used to mean not very good, rather than covered in dung.

Soil, possibly from Old French *soiller*, from the Latin *sucula*, diminutive of *sus*, pig, perhaps the scurrilous implication being of pigs wallowing in their own excrement. A general term for waste, sewage, excrement. Now rather archaic, and apart from 'soil pipe', the large-diameter drain from the lavatory pan, barely used in modern language. Related to 'sullage', previously used for sewage in general, but recently more for waste water only. Separate from 'soil' in the sense of the earthen layer in which plants grow. The two are confused in the notion of something being 'soiled' (dirtied), which seems, to me, closer to the excremental than the horticultural. See 'night soil'.

Spend a penny, quaint euphemism for going to the toilet. The penny was used in the pay-slot on the cubicle doors of public toilets, a charge to cover their upkeep. Often used to imply urination, on the grounds, perhaps, that a penny was not very much – a small thing – but if my memory serves, gent's toilets at least were free to use the urinals. So the charge was actually only applied to something bigger. Nothing to do with 'pee'; though in the UK the modern penny (100 = £1) is abbreviated to 'p', the pennies spent in public lavatories were invariably the older and much larger penny (240 = £1), which was designated 'd', from the Roman coin *denarius*.

Spoor, from the Afrikaans *spoor*, originally Dutch (sometimes *spor*), meaning tracks. Traditionally used in hunting to mean signs, scent, footprints or droppings of an animal that a tracker can use to follow it during the hunt or safari trip. Incidentally, if you do a Google image search for 'spoor' you don't get footprints or droppings, you get hundreds of pictures of railway tracks, courtesy of Dutch transport and engineering websites.

Spraints, from Old French *espreintes*, from *espraindre*, to squeeze out, after the Latin *exprimere* to express. Otter excrement, which is squeezed out, along with a copious secretion of anal jelly lubricant, to aid the passage of sharp fish bones. Used mainly by hunters and naturalists.

Stallage, originally from the French *estallage*, itself from *estal*, a stall (in the sense of a stable) and meaning both the rent or right for creating, keeping or using a stall to keep an animal, and subsequently the dung dropped whilst the stall was in use. Perhaps echoing words like spillage and sewage.

Stercore, Latin for dung. Hardly used nowadays, but there was a fashion amongst some entomologists to record dung beetles as being found in *stercore equino* (horse dung), *stercore ovino* (sheep) or *stercore bovino* (cattle). Latin descriptions and accounts were once intermingled with English, but this prudish excuse for not using the word 'dung' occurred well into the 20th century. There was also the occasional reference to the more adventurous entomologists finding beetles in *stercore humano*. Also gives us 'stercovorous' eating dung, 'stercorate' dung or manure, 'stercorary' a dung or manure store, 'stercoricolous' for a plant growing out of dung, 'stercorite' a mineral found in guano, and 'Sterculius' a Roman god who presided over manuring. The yellow dung fly is *Scathophaga stercoraria*. The skuas, large seabirds also called jaegers, have the generic name *Stercorarius*, from their habit of eating carrion and food they have frightened other birds into vomiting up.

Stool, from Dutch *stoel* and German *Stuhl*, a chair. A sample of human faeces. Originally a real stool, seat or chair with a hole, on which to sit whilst defecating into a pot underneath, but later to mean the bowel evacuation itself. Now only used in medical texts, hence the Bristol Stool Chart, a visual guide to textural consistency of stools, used in academic studies and medical diagnosis.

Tath, also **tad**, **taith** or **teathe**, from Norse *tath*, dung and *tatha*, a manured field. Highly obscure and dialectical, Scottish? As well as meaning the dung, or manure itself, it can also mean the strong grass growing round the decomposing remains of cattle dung. *Tadfall* self-explanatory. Probably not used in any agricultural textbook this side of 1850.

Toilet, from French *toilette*, a cloth, diminutive of *toile*, and originally a place to wash and/or dress, the cloth being a cover for a dressing table, or a towel. Now meaning the room, or the plumbed bowl, where urination and defecation takes place. See 'lavatory'. One of my old school teachers refused to allow us to leave the class if we asked to go to the toilet; she insisted we asked to go to the lavatory, as if this were in some way a more refined term. I never liked her.

Treddles, and **trottles**, seemingly similar to treadle, from Anglo Saxon *tredel*, tread. Archaic term for dung of sheep or hare, something to do with them being trodden on? It's tempting to imagine that 'the trots', euphemistically informal for an attack of diarrhoea, might have some connection here, but maybe that use is just as likely to be an implication of speed, either of the output, or of the victim moving towards the toilet.

Turd, from old English *tord*, excrement. Now used only in vulgar slang. In Old English, a dung beetle was a *tordwifel* (literally a turd weevil), and it is still *tordivel* in modern Norwegian. According to some sources it is represented in cockney rhyming slang as a Richard (the Third). This being my name I prefer Richard to mean 'bird'. This fits much better with the old-fashioned shortening of Richard, through Ricky, to Dicky, thus also the much more familiar 'dicky bird' = 'word', as in: 'Don't say a dicky bird about this faecal association'. Got that?

Urine, from the Latin *urina*. Watery liquid discharged from the bladder, containing urea, $CO(NH_2)_2$, a safe, non-toxic waste product to remove ammonia (NH_3) after protein metabolism. In birds and other non-mammalian animals, nitrogen is excreted as a heterocyclic compound, $C_5H_4N_4O_3$, uric acid.

Waggyings, etymology uncertain. Reputedly the dung of the fox, according to ancient treatises on rural economy, wildlife and hunting.

Wee, childish word for urine. Also 'wee-wee'. See 'pee'.

Werdrobe, or **Werderobe**, so far I'm unable to find anything about the etymology of this word. According to various very old, very obscure and very questionable books, it is the dung of the badger. Nowadays, as any internet search will quickly show, it is all too frequently a spelling mistake for 'wardrobe'.

Water closet, a plumbed toilet, using a water flush, and a bit more advanced than an earth closet. Nowadays more often known by its initials WC.

REFERENCES

Anderson, J.M and Coe, M.J. (1974) Deposition of elephant dung in an arid tropical environment. *Oecologia* 14: 111–125.

Arillo, A. and Ortuno, V.M. (2008) Did dinosaurs have any relation with dung-beetles? (The origin of coprophagy). *Journal of Natural History* 42: 1405–1408.

Barwise, S. (1904) *The Purification of Sewage, being a Brief Account of the Scientific Principles of Sewage Purification and their Practical Application.* London: Crosby Lockwood.

Bates, W.H. (1886–1890) *Biologia Centrali-Americana. Insecta. Coleoptera. Pectinicornia and Lamellicornia.* Vol. 2, part 2. London: R.H. Porter.

Beaune, D., Bollache, L., Bretagnolle, F and Fruth, B. (2012) Dung beetles are critical in preventing post-dispersal seed removal by rodents in Congo rain forest. *Journal of Tropical Ecology* 28: 507–510.

Berry, P. (1993) From cow pat to frying pan: Australian herring (*Arripes georgianus*) feed on an introduced dung beetle (Scarabaeidae). *Western Australian Naturalist* 19: 241–242.

Bewick, T. (1790) *A General History of Quadrupeds.* Newcastle upon Tyne: Hodgson, Beilby and Bewick.

Beynon, S.A. (2012) Potential environmental consequences of administration of antihelminthics to sheep. *Veterinary Parasitology* 189: 113–124.

Bowie, G.G.S. (1987) New sheep for old – changes in sheep farming in Hampshire, 1792–1879. *Agricultural History Review* 35: 15–24.

Bradley, J.D. (1982) Two new species of moths (Lepidoptera, Pyralidae, Chrysauginae) associated with the three-toed sloth (*Bradypus* spp.) in South America. *Acta Amazonica* 12: 649–656.

Bragg, A.N. (1957) Use of carrion by the beetle *Canthon laevis* (Coleoptera, Scarabaeidae). *Southwestern Naturalist* 2: 173.

Brussaard, L. (1983) Reproductive behaviour and development of the dung beetle *Typhaeus typhoeus* (Coleoptera: Geotrupidae). *Tijdschrift voor Entomologie* 126: 203–231.

Buckton, G.B. (1895) *The Natural History of* Eristalis tenax *or the Drone-fly.* London: Macmillan.

Cambefort, Y. and Walter, P. (1985) Description du nid et de la larve de *Paraphytus aphodioides* Boucomont et notes sur l'origine de la coprophagie et l'évolution des Coléoptères Scarabaeidae s. str. *Annales de la Société Entomologique de France (NS)* 21: 351–356.

Carpaneto, G.M., Mazziotta, A. and Valerio, L. (2007) Inferring species decline from collection records: roller dung beetles in Italy (Coleoptera, Scarabaeidae). *Diversity and Distributions* 13: 903–919.

Chapman, T.A. (1869) *Aphodius porcus*, a cuckoo parasite on *Geotrupes stercorarius*. *Entomologist's Monthly Magazine* 5: 273–276.

Cheyne, G. (1715) *Philosophical Principles of Religion, Natural and Revealed, in Two Parts. Part 1, Containing the Elements of Natural Philosophy and the Proofs of Natural Religion*. London: George Strahan.

Coe, M. (1977) The role of termites in the removal of elephant dung in the Tsavo (East) National Park Kenya. *East African Journal of Wildlife* 49: 49–55.

Coggan, N. (2012) Are native dung beetle species following mammals in the critical weight range towards extinction? *Proceedings of the Linnean Society of New South Wales* 134: A5–A9.

Constantine, B. (1994) A new ecological niche for *Gibbium aequinoctiale* Boieldieu (Ptinidae) in Britain, and a reconsideration of literature references to *Gibbium* spp. *The Coleopterist* 3: 25–28.

Coope, G.R. (1973) Tibetan species of dung beetle from late Pleistocene deposits in England. *Nature* 245: 335–336.

Cullen, P and Jones, R. (2012) Manure and middens in English place-names. In R. Jones (ed.) *Manure Matters: Historical, Archaeology and Ethnographic Perspectives*. Leicester: Ashgate.

Cummings, J.H. (1984) Constipation, dietary fibre and the control of large bowel function. *Postgraduate Medical Journal* 60: 811–819.

Curtis, J. (1823–1840) *British Entomology: Being Illustrations and Descriptions of the Genera of Insects found in Great Britain and Ireland... etc.* London: Printed for the author.

Curtis, V., Aunger, R. and Rabie, T. (2004) Evidence that disgust evolved to protect from risk of disease. *Proceedings of the Royal Society B (Suppl.)* 271: S131–S133.

Dacke, M., Baird, E., Byrne, M., Scholtz, C.H. and Warrant, E.J. (2013) Dung beetles use the Milky Way for orientation. *Current Biology* 23: 298–300.

Dalgleish, E.A. and Elgar, M.A. (2005) Breeding ecology of the rainforest dung beetle *Cephalodesmius armiger* (Scarabaeidae) in Tooloom National Park. *Australian Journal of Zoology* 53: 95–102.

Darby, M. (2014) Pitfall trap surveys of beetles in Langley Wood National Nature Reserve, Wiltshire. *British Journal of Entomology and Natural History* 27: 27–43.

Darwin, C. (1839) *Journal of Researches into the Geology and Natural History of the Various Countries Visited by HMS Beagle under the Command of Captain Fitzroy, RN from 1832 to 1836*. London: Henry Colburn.

Darwin, C. (1871) *The Descent of Man, and Selection in Relation to Sex*. London: John Murray.

Darwin, C. (1881) *The Formation of Vegetable Mould through the Action of Worms with Observations on their Habits*. London: John Murray.

Dennis, R.W.G. (1960) *British Cup Fungi and their Allies: an Introduction to the Ascomycetes*. London: Ray Society.

Disney, R.H.L. (1974) Speculations regarding the mode of evolution of some remarkable associations between Diptera (Cuterebridae, Simuliidae and Sphaeroceridae) and other arthropods. *Entomologist's Monthly Magazine* 110: 67–74.

Dortel, E., Thuiller, W., Lobo, J.M., Bohbot, H., Lumaret, J.P. and Jay-Robert, P. (2013) Potential effects of climate change on the distribution of Scarabaeidae dung beetles in Western Europe. *Journal of Insect Conservation* 17: 1059–1070.

Doube, B.M., Macqueen, A., Ridsdill-Smith, T.J. and Weir, T.A. (1991) Native and introduced dung beetles in Australia. In I. Hanski and Y. Cambefort (eds) *Dung Beetle Ecology*. Princeton: Princeton University Press. pp. 255–278.

Edwards, P.B. (2007) *Introduced Dung Beetles in Australia 1967–2007 – Current Status and Future Directions*. Maleny, Queensland: Dung Beetles for Landcare Farming Community.

Eisner, T. and Eisner, M. (2000) Defensive use of a fecal thatch by a beetle larva (*Hemisphaerota cyanea*). *Proceedings of the National Academy of Science of the United States of America* 97: 2632–2636.

el Jundi, B., Foster, J.J., Khaldy, L., Byrne, M.J., Dacke, M. and Baird, E. (in press) A snapshot-based mechanism for celestial orientation. *Current Biology*.

Fabre, J.-H. (1897) *Souvenirs entomologiques*. Paris: Delagrave.

Fabre, J.H. (1921) *Fabre's Book of Insects. Retold by Mrs Rodolph Stawall, illustrated by E.J. Detmold*. London: Hodder & Stoughton.

Farnworth, E.R., Modler, H.W. and Mackie, D.A. (1995) Adding Jerusalem artichoke (*Helianthus tuberosus* L.) to weanling pig diets and the effect on manure composition and characteristics. *Animal Feed Science and Technology* 55: 153–160.

Fenolio, D.B., Graening, G.O., Collier, B.A. and Stout, J.F. (2006) Coprophagy in a cave-adapted salamander; the importance of bat guano examined through nutritional and stable isotope analyses. *Proceedings of the Royal Society B* 273: 439–443.

Fijen, T.P.M., Kamp, J., Lameris, T.K., Pulikova, G., Urazaliev, R., Kleijn, D.

and Donald, P.F. (2015) Functions of extensive animal dung 'pavements' around the nests of the black lark (*Melanocorphya yeltoniensis*). *The Auk* 132: 878–892.

Floate, K.D. (2011) Arthropods in cattle dung on Canada's grasslands. In D.J. Floate (ed.) *Anthropods of Canadian grasslands*, Vol. 2: *Inhabitants of a Changing Landscape*. Biological Survey of Canada. pp. 71–88.

Fowler, W.W. (1890) *The Coleoptera of the British Islands. A Descriptive Account of the Families, genera and Species...etc. etc.* London: L. Reeve. Vol. 4.

Freymann, B., Buitenwerf, R., Desouza, O. and Olff, H. (2008) The importance of termites (Isoptera) for the recycling of herbivore dung in tropical ecosystems: a review. *European Journal of Entomology* 105: 165–173.

Galloway, J.M., Adamczewski, J., Schock, D.M., Andrews, T.D., MacKay, G., Bowyer, V.E., Meulendyk, T., Moorman, B.J. and Kutz, S.J. (2012) Diet and habitat of mountain woodland caribou inferred from dung preserved in 5000-year-old alpine ice in Swleyn Mountains, Northwest Territories, Canada. *Arctic* 65 (Suppl. 1): 59–79.

Goodhart, J.F. (1902) Round about constipation. *Lancet* ii: 1241.

Grebennikof, V.V. and Scholtz, C.H. (2004) The basal phylogeny of Scarabaeoidea (Insecta: Coleoptera) inferred from larval morphology. *Invertebrate Systematics* 18: 321–348.

Grimaldi, D. and Engel, M.S. (2005) *Evolution of the Insects*. Cambridge: Cambridge University Press.

Gunter, N.L., Weir, T.A., Slipinksi, A., Bocak, L. and Cameron, S.L. (2016) If dung beetles (Scarabaeidae: Scarabaeinae) arose in association with dinosaurs, did they also suffer a mass co-extinction at the K–Pg boundary? *PLoS ONE* 11(5): e0153570.

Halffter, G., Halffter, V and Favila, M.E. (2011) Food relocation and the nesting behavior in *Scarabaeus* and *Kheper* (Coleoptera: Scarabaeinae). *Acta Zoológica Mexicana (NS)* 27: 305–324.

Hanski, I. and Cambefort, Y. (eds) (1991) *Dung Beetle Ecology*. Princeton: Princeton University Press.

Harvey, P.H. and Godfray, C.J. (2001) A horn for an eye. *Science* 291: 1505–1506.

Heinrich, B. and Bartholomew, G.A. (1979a) Roles of endothermy and size in inter- and infra-specific competition for elephant dung in an African dung beetle, *Scarabaeus laevistriatus*. *Physiological Zoology* 52: 484–496.

Heinrich, B. and Bartholomew, G.A. (1979b) The ecology of the African dung beetle. *Scientific American* 241(5): 146–156.

Hertel, F. and Colli, G.R. (1998) The use of leaf-cutter ants, *Atta laevigata* (Smith) (Hymenoptera: Formicidae), as a substrate for oviposition by the dung beetle *Canthon virens* Mannerheim (Coleoptera: Scarabaeidae) in central Brazil. *Coleopterists Bulletin* 52: 105–108.

Hewitt, C.G. (1914) *The house fly,* Musca domestica, *Linnaeus. A Study of its Structure, Development, Bionomics and Economy*. Manchester: University of Manchester Press.

Hodge, P.J. (1995) *Copris lunaris* (L.) (Scarabaeidae) in Sussex. *The Coleopterist* 3: 82–83.

Hogue, C.L. (1983) An entomological explanation of Ezekial's wheels? *Entomological News* 94: 73–80.

Holloway, B.A. (1976) A new bat-fly family from New Zealand (Diptera: Mystacinobiidae). *New Zealand Journal of Zoology* 3: 279–301.

Holter, P. and Scholtz, C.H. (2007) What do dung beetles eat? *Ecological Entomology* 32: 690–697.

Howard, L.O. (1900) A contribution to the study of the insect fauna of human excrement (with especial reference to the spread of typhoid fever by flies). *Proceedings of the Washington Academy of Sciences* 2: 541–604.

Howden, H. (1952) A new name for *Geotrupes* (*Peltotrupes*) *chalybaeus* LeConte with a description of the larvae and its biology. *Coleopterists Bulletin* 6: 41–48.

Janzen, D.H. (1986) Mice, big mammals, and seeds: it matters who defecates where. In A. Estrada and T.E. Fleming (eds) *Frugivores and Seed Dispersal.* Dordrecht: Junk. pp. 251–272.

Jones, A.W. (1961) The vegetation of the South Norwood or Elmers End Sewage Works. *London Naturalist* 40: 102–114.

Jones, R.A. (1984) *Vespula germanica* (F.) wasps hunting dung beetles *Aphodius contaminatus* (L.). *Proceedings and Transactions of the British Entomological and Natural History Society* 17: 36–37.

Jones, R.A. (1986) Some novel collecting methods for the coleopterist. *Bulletin of the Amateur Entomologists' Society* 45: 21–24.

King, F.H. (1911) *Farmers of Forty Centuries, or Permanent Agriculture in China, Korea and Japan.* Madison, WI: Mrs F.H. King.

Kirk-Spriggs, A.H., Kotrba, M and Copeland, R.S. (2011) Further details of the morphology of the enigmatic African fly *Mormotomyia hirsuta* Austen (Diptera: Mormotomyiidae). *African Invertebrates* 52: 145–165.

Klausnitzer, B. (1981) *Beetles.* New York: Exeter Books.

Knell, R. (2011) Male contest competition and the evolution of weapons. In L.W Simmons and T.J. Ridsdill-Smith (eds) *Ecology and Evolution of Dung Beetles.* Oxford: Wiley-Blackwell. pp. 47–65.

Koskela, H. and Hanski, I. (1977). Structure and succession in a beetle community inhabiting cow dung. *Annales Zoologici Fennici* 14: 204–223.

Larsen, T.H., Lopera, A., Forsyth, A. and Génier, F. (2009) From coprophagy to predation: a dung beetle that kills millipedes. *Biology Letters* 5: 152–155.

Lavoie, K.H., Helf, K.L and Poulson, T.L. (2007) The biology and ecology of North American cave crickets. *Journal of Cave and Karst Studies* 69: 114–134.

Lobo, J.M. (2001) Decline of roller dung beetle (Scarabaeinae) populations in the Iberian peninsula during the 20th century. *Biological Conservation* 97: 43–50.

Lumaret, J.-P. (1986) Toxicité de certains helminthicides vis-a-vis des insectes coprophages et conséquences sur la disparition des excréments de la surface du sol. *Acta Oecologica Oecologia Applicata* 7: 313–324.

Martin, A.J. (1935) *The Work of the Sanitary Engineer. A Handbook for Engineers, Students and Others Concerned with Public Health.* London: MacDonald and Evans.

Maruyama, M. (2012) A new genus and species of flightless, microphthalmic Corythoderini (Coleoptera: Scarabaeidae: Aphodiinae) from Cambodia, associated with *Macrotermes* termites. *Zootaxa* 3555: 83–88.

Matthews, E.G. (1963) Observations on the ball-rolling behaviour of *Canthon pilularius* (L.) (Coleoptera: Scarabaeidae). *Psyche* 70: 75–93.

McConnell, P. (1883) *Note-book of Agricultural Facts and Figures for Farmers and Farm Students.* London: MacDonald and Martin.

Medina, C.A., Molano, F. and Scholtz, C.H. (2013) Morphology and terminology of dung beetles (Coleoptera: Scarabaeidae: Scarabaeinae) male genitalia. *Zootaxa* 3626: 455–476.

Michelet, J. (1875) *The Insect.* London: Nelson & Sons.

Midgley, J.J., White, J.D.M., Johnson, S.D. and Bronner, G.N. (2015) Faecal mimicry by seeds ensures dispersal by dung beetles. *Nature Plants* 1: 15141.

Moczec, A.P. (2006) Integrating micro- and macro-evolution of development through the study of horned beetles. *Heredity* 97: 168–178.

Moczec, A.P. and Nijhout, H.F. (2004) Trade-offs during the development of primary and secondary sexual traits in a horned beetle. *American Naturalist* 163: 184–191.

Negro, J.J., Grande, J.M., Tella, J.L., Garrido, J., Hornero, D., Donazar, J.A., Sanchez-Zapata, J.A., Benitez, J.R. and Barcell, M. (2002) An unusual source of essential carotenoids. A yellow-faced vulture includes ungulate faeces in its diet for cosmetic purposes. *Nature* 416: 807.

Nichols, E. and Gómez, A. (2014) Dung beetles and fecal helminth transmission: patterns, mechanisms and questions. *Parasitology* 141: 614–623.

Nichols, E., Gardner, T.A., Peres, C.A., Spector, S. and the Scarabaeinae Research Network (2009) Co-declining mammals and dung beetles: an impending ecological cascade. *Oikos* 118: 481–487.

Nilssen, A.C., Åsbakk, K., Haugerus, R.E., Hemmingsen, W and Oksanen, A. (1999) Treatment of reindeer with ivermectin – effects on dung insect fauna. *Rangifer* 19: 61–69.

Parker, G.A. (1970) Sperm competition and its evolutionary effect on copula duration in the fly *Scatophaga stercoraria. Journal of Insect Physiology* 16: 1301–1328.

Pauli, J.N., Mendoza, J.E., Steffan, S.A., Carey, C.C., Weimer, P.J. and Peery, M.Z. (2014) A syndrome of mutualism reinforces the lifestyle of a sloth. *Proceedings of the Royal Society B* 281: 20133006.

Peck, S.B. and Kukalova-Peck, J. (1989) Beetles (Coleoptera) of an oil-bird cave: Cueva del Guacharo, Venezuela. *Coleopterists Bulletin* 43: 151–156.

Pérèz-Ramos, I.M., Marañon, T., Lobo, J.M. and Verdu, J.R. (2007) Acorn removal and dispersal by the dung beetle *Thorectes lusitanicus*: ecological implications. *Ecological Entomology* 32: 349–356.

Philips, T.K. (2011) The evolutionary history and diversification of dung beetles. In L.W Simmons and T.J. Ridsdill-Smith (eds), *Ecology and Evolution of Dung Beetles.* Oxford: Wiley-Blackwell. pp. 21–46.

Pizo, M.A., Guimarães, P.R. and Oliviera, P.S. (2005) Seed removal by ants produced by different vertebrate species. *Ecoscience* 12: 136–140.

Popp, J.W. (1988) Selection of horse dung pats by foraging house sparrows. *Field Journal of Ornithology* 59: 385–388.

Pratt, T.K. (1988). *Dictionary of Prince Edward Island English.* Toronto: University of Toronto Press.

Putman, R.J. (1983) *Carrion and Dung. The Decomposition of Animal Wastes.* Studies in Biology No. 156. London: Edward Arnold.

Ratcliffe, B.C. (1980) A new species of Coprini (Coleoptera: Scarabaeidae: Scarabaeinae) taken from the pelage of three toed sloths (*Bradypus tridactylus* L.) (Edentata: Bradypodidae) in central Amazonia with a brief commentary on scarab–sloth relationships. *Coleopterists Bulletin* 34: 337–350.

Reitter, E. 1908–1916. *Fauna Germanica. Die Käfer des Deutschen Reiches.* Stuttgart: Lutz, 5 vols.

Richardson, M.J. and Watling, R. (1997) *Keys to Fungi on Dung.* Stourbridge: British Mycological Society.

Rideal, S. (1900) *Sewage and the Bacterial Purification of Sewage.* London: Robert Ingram.

Ridsdill-Smith, T.J. and Edwards, P.B. (2011) Biological control: ecosystem functions provided by dung beetles. In L.W Simmons and T.J. Ridsdill-Smith (eds) *Ecology and Evolution of Dung Beetles.* Oxford: Wiley-Blackwell. pp. 245–266.

Sajo, K. (1910) *Aus dem Leben der Käfer.* Leipzig: Thomas.

Salgado, S.S., Motta, P.C., de Souza Aguiar, L.M. and Nardoto, G.B. (2014) Tracking dietary habits of cave arthropods associated with deposits of hematophagous bat guano: a study from a neotropical savanna. *Austral Ecology* 39: 560–566.

Salmon, W. (1693) *Seplasium. The Compleat English Physician: or the Druggist's Shop Opened, Explicating all the Particulars of which Medicines at this Day are Composed and Made, Shewing their Various Names and Natures.* London: Gilliflower and Sawbridge.

Sánchez, M.V. and Genise, G.F. (2009) Cleptoparasitism and detritivory in dung beetle fossil brood balls from Pategonia, Argentina. *Palaeontology* 52: 837–848.

Scholtz, C.H., Davis, A. and Kryger, U. (eds) (2009) *Evolutionary Biology and Conservation of Dung Beetles.* Sofia: Pensoft Publishers.

Scholtz, C.H., Harrison, J.du G. and Grebennikov, V.V. (2004) Dung beetle (*Scarabaeus* (*Pachysoma*)) biology and immature stages: reversal to ancestral states under desert conditions Coleoptera: Scarabaeidae)? *Biological Journal of the Linnean Society* 83: 453–460.

Shaw, G. (1806) *General Zoology or Systematic Natural History.* Vol. VI, Part II, Insects. London: Kearsley.

Shepherd, V.A. and Chapman, C.A. (1998) Dung beetles as secondary seed dispersers: impact on seed predation and germination. *Journal of Tropical Ecology* 14: 199–215.

Simmons, L.W. and Emlen, D.J. (2006) Evolutionary trade-off between weapons and testes. *Proceedings of the National Academy of Sciences of the United States of America* 103: 16346–16351.

Simmons, L.W. and Ridsdill-Smith, T.J. (eds) (2011a) *Ecology and Evolution of Dung Beetles.* Oxford: Wiley-Blackwell.

Simmons, L.W. and Ridsdill-Smith, T.J. (2011b) Reproductive competition and its impact on the evolution and ecology of dung beetles. In L.W Simmons and T.J. Ridsdill-Smith (eds) *Ecology and Evolution of Dung Beetles.* Oxford: Wiley-Blackwell. pp. 1–20.

Skidmore, P. (1991) *Insects of the British Cow-dung Community.* Occasional Publication No. 21. Preston Montford: Field Studies Council.

Smith, A.B.T., Hawks, D.C. and Heraty, J.M. (2006) An overview of the classification and evolution of the major scarab beetle clades (Coleoptera: Scarabaeoidea) based on preliminary molecular analyses. *Coleopterists Society Monographs* 5: 35–46.

Stavert, J.R., Gaskett, A.C., Scott, D.J. and Beggs, J.R. 2014. Dung beetles in an avian-dominated island ecosystem: feeding and trophic ecology. *Oecologia* 176: 259–271.

Sutton, G., Bennett, J. and Bateman, M. (2013) Effects of ivermectin residues on dung invertebrate communities in a UK farmland habitat. *Insect Conservation and Diversity* 7: 64–72.

Swammerdam, J. (1669) *Historia Insectorum Generalis, etc.* Utrecht: Meinardus van Drevnen.

Sykes, W.H. (1835) Observations upon the habits of *Copris midas*. *Transactions of the Entomological Society of London* 1: 130–132.

Telfer, M.G., Lee, P. and Lyons, G. (2004) The pride of Kent *Emus hirtus* (L. 1758) at Elmley Marshes RSPB reserve. *Bulletin of the Amateur Entomologists' Society* 63: 44–46.

Tribe, G.D and Burger, B.V. (2011) Olfactory ecology. In L.W Simmons and T.J. Ridsdill-Smith (eds) *Ecology and Evolution of Dung Beetles*. Oxford: Wiley-Blackwell. pp. 87–106.

Vaz-de-Mello, F.Z. (2007) Revision and phylogeny of the dung beetle genus *Zonocopris* Arrow 1932 (Coleoptera: Scarabaeidae: Scarabaeinae), a phoretic of land snails. *Annales de la Société Entomologique de France (NS)* 43: 231–239.

Wall, R. and Strong, L. (1987) Environmental consequences of treating cattle with the antiparasitic drug ivermectin. *Nature* 327: 418–421.

Wallace, A.R. (1853) *Narrative of Travels on the Amazon and Rio Negro, with an Account of the Native Tribes, and Observations on the Climate, Geology, and Natural History of the Amazon Valley*. London: Reeve & Co.

Wassell, J.L.H. (1966) Coprophagous weevils (Coleoptera: Curculionidae). *Australian Journal of Entomology* 5: 73–74.

Young, O.P. (1978) *Resource Partitioning in a Neotropical Necophagous Scarab Guild*. PhD thesis, University of Maryland.

Young, O.P. (1981a) The attraction of Neotropical Scarabaeinae (Coleoptera: Scarabaeidae) to reptile and amphibian fecal material. *Coleopterists Bulletin* 35: 345–348.

Young, O.P. (1981b) The utilization of sloth dung in a neotropical forest. *Coleopterists Bulletin* 35: 427–430.

INDEX